Tucholsky Wagner Zola Scott Sydow Freud Schlegel
Turgenev Wallace Fonatne
Twain Walther von der Vogelweide Fouqué Friedrich II. von Preußen
Weber Freiligrath Frey
Fechner Fichte Weiße Rose von Fallersleben Kant Ernst Frommel
Hölderlin Richthofen
Engels Fielding Eichendorff Tacitus Dumas
Fehrs Faber Flaubert
Eliasberg Ebner Eschenbach
Feuerbach Maximilian I. von Habsburg Fock Zweig
Eliot Vergil
Ewald
Goethe Elisabeth von Österreich London
Mendelssohn Balzac Shakespeare Dostojewski Ganghofer
Lichtenberg Rathenau Doyle Gjellerup
Trackl Stevenson Hambruch
Mommsen Tolstoi Lenz Droste-Hülshoff
Thoma Hanrieder
von Arnim Hägele Humboldt
Dach Verne Hauff
Reuter Rousseau Hagen Hauptmann Gautier
Karrillon Garschin
Defoe Baudelaire
Damaschke Hebbel
Descartes
Hegel Kussmaul Herder
Wolfram von Eschenbach Dickens Schopenhauer
Darwin Melville Grimm Jerome Rilke George
Bronner
Campe Horváth Aristoteles Bebel Proust
Bismarck Vigny Barlach Voltaire Federer Herodot
Gengenbach Heine
Storm Casanova Tersteegen Grillparzer Georgy
Chamberlain Lessing Langbein Gilm
Gryphius
Brentano Lafontaine
Strachwitz Claudius Schiller Kralik Iffland Sokrates
Bellamy Schilling
Katharina II. von Rußland
Gerstäcker Raabe Gibbon Tschechow
Löns Hesse Hoffmann Gogol Wilde Vulpius
Gleim
Luther Heym Hofmannsthal Morgenstern
Klee Hölty Goedicke
Roth Heyse Klopstock Kleist
Luxemburg Puschkin Homer Mörike
La Roche Horaz Musil
Machiavelli
Navarra Aurel Musset Kierkegaard Kraft Kraus
Lamprecht Kind Kirchhoff Hugo Moltke
Nestroy Marie de France
Laotse Ipsen Liebknecht
Nietzsche Nansen
Marx Ringelnatz
von Ossietzky Lassalle Gorki Klett Leibniz
May
vom Stein Lawrence Irving
Petalozzi
Platon Knigge
Sachs Pückler Michelangelo Kock Kafka
Poe Liebermann
de Sade Praetorius Mistral Zetkin Korolenko

The publishing house tredition has created the series **TREDITION CLASSICS**. It contains classical literature works from over two thousand years. Most of these titles have been out of print and off the bookstore shelves for decades.

The book series is intended to preserve the cultural legacy and to promote the timeless works of classical literature. As a reader of a **TREDITION CLASSICS** book, the reader supports the mission to save many of the amazing works of world literature from oblivion.

The symbol of **TREDITION CLASSICS** is Johannes Gutenberg (1400 – 1468), the inventor of movable type printing.

With the series, tredition intends to make thousands of international literature classics available in printed format again – worldwide.

All books are available at book retailers worldwide in paperback and in hardcover. For more information please visit: www.tredition.com

tredition was established in 2006 by Sandra Latusseck and Soenke Schulz. Based in Hamburg, Germany, tredition offers publishing solutions to authors and publishing houses, combined with worldwide distribution of printed and digital book content. tredition is uniquely positioned to enable authors and publishing houses to create books on their own terms and without conventional manufacturing risks.

For more information please visit: www.tredition.com

Essays Political, Economical, and Philosophical —Volume 1

Benjamin, Graf von Rumford

Imprint

This book is part of the TREDITION CLASSICS series.

Author: Benjamin, Graf von Rumford
Cover design: toepferschumann, Berlin (Germany)

Publisher: tredition GmbH, Hamburg (Germany)
ISBN: 978-3-8495-1334-4

www.tredition.com
www.tredition.de

Copyright:
The content of this book is sourced from the public domain.

The intention of the TREDITION CLASSICS series is to make world literature in the public domain available in printed format. Literary enthusiasts and organizations worldwide have scanned and digitally edited the original texts. tredition has subsequently formatted and redesigned the content into a modern reading layout. Therefore, we cannot guarantee the exact reproduction of the original format of a particular historic edition. Please also note that no modifications have been made to the spelling, therefore it may differ from the orthography used today.

VOL. I.

Contents

Dedication

Dedication

To his most serene highness THE ELECTOR PALATINE reigning duke of bavaria. etc. etc. etc.

SIR,

In requesting permission to dedicate to you most Serene Electoral Highness these Essays, I had several important objects in view: I was desirous of showing to the world that I had not presumed to publish an account of public measures and institutions, planned and executed in your Electorial Highness's dominions,—by your orders,—and under your immediate authority and protection, without your leave and approbation. I was also desirous of availing myself of the illustrious name of a Sovereign eminently distinguished by his munificence in promoting useful knowledge, and by his solicitude for the happiness and prosperity of his subjects, to recommend the important objects I have undertaken to investigate, to the attention of the Great,—the Wise,—and the Benevolent. And lastly, I was anxious to have an opportunity of testifying, in a public manner, my gratitude to your most Serene Electoral Highness for all your kindness to me; and more especially for the distinguished honour you have done me by selecting and employing me as an instrument in your hands of doing good.

I have the honour to be, with the most profound respect, and with unalterable attachment,

SIR,
Your Most Serene ELECTORIAL HIGHNESS's

Devoted Servant,

RUMFORD.

London,

July, 1st, 1796.

CONTENTS of ESSAY I.

an ACCOUNT of an ESTABLISHMENT FOR THE POOR AT MUNICH

together with

A Detail of various Public Measures, connected with that Institution, which have been adopted and carried into effect for putting an End to Mendicity, and introducing Order, and useful Industry, among the more Indigent of the Inhabitants of Bavaria.

Introduction

CHAPTER. I. Of the prevalence of mendicity in Bavaria at the time when the measures for putting an end to it were adopted.

CHAPTER. II. Various preparations made for putting an end to mendicity in bavaria. Cantonment of the cavalry in the country towns and villages. Formation of the committee placed at the head of the institution for the poor at Munich. The funds of that institution.

CHAPTER. III. Preparations made for giving employment to the poor. Difficulties attending that undertaking. The measures adopted completely successful. The poor reclaimed to habits of useful industry. Description of the house of industry at Munich.

CHAPTER. IV. An account of the taking up of the beggars at Munich. The inhabitants are called upon for their assistance. General subscription for the relief and support of the poor. All other public and private collections for the poor abolished.

CHAPTER. V. The different kinds of employment given to the beggars upon their being assembled in the house of industry. Their great awkwardness at first. Their docility, and their progress in useful industry. The manner in which they were treated.

The manner in which they were fed. The Precautions used to prevent Abuses in the Public Kitchen from which they were fed.

CHAPTER. VI. Apology for the want of method in treating the subject under consideration. Of the various means used for encouraging industry among the poor. Of the internal arrangement and government of the house of industry. Why called the military work-house. Of the manner in which the business is carried on there. Of the various means used for preventing frauds in carrying on the business in the different manufactures. Of the flourishing state of those manufactures.

CHAPTER. VII. A further account of the poor who were brought together in the house of industry:—and of the interesting change which was produced in their manners and dispositions. Various proofs that the means used for making them industrious, comfortable, and happy, were successful.

CHAPTER. VIII. Of the means used for the relief of those poor persons who were not beggars. Of the large sums of money distributed to the poor in alms. Of the means used for rendering those who received alms industrious. Of the general utility of the house of industry to the poor, and the distressed of all denominations. Of public kitchens for feeding the poor, united with establishments for giving them employment; and of the great advantages which would be derived from forming them in every parish. Of the manner in which the poor of Munich are lodged.

CHAPTER. IX. Of the means used for extending the influence of the institution for the poor at Munich, to other parts of Bavaria. Of the progress which some of the improvements introduced at Munich are making in other countries.

INTRODUCTION.

[IMAGE] view of the Military Workhouse at Munich

Situation of the Author in the Service of His Most Serene
Highness the ELECTOR PALATINE, Reigning Duke of BAVARIA.
Reasons which induced him to undertake to form an Establishment
for the Relief of the Poor.

Among the vicissitudes of a life chequered by a great variety of
incidents, and in which I have been called upon to act in many interesting scenes, I have had an opportunity of employing my attention upon a subject of great importance; a subject intimately and
inseparably connected with the happiness and well-being of all civil
societies; and which, from its nature, cannot fail to interest every
benevolent mind;—it is the providing for the wants of the Poor, and
the securing their happiness and comfort by the introduction of
order and industry among them.

The subject, though it is so highly interesting to mankind, has not
yet been investigated with that success that could have been
wished. This fact is apparent, not only from the prevalence of indolence, misery, and beggary, in almost all the countries of Europe;
but also from the great variety of opinion among those who have
taken the matter into serious consideration, and have proposed
methods for remedying those evils; so generally, and so justly complained of.

What I have to offer upon the this subject being not merely specu-
lative opinion, but the genuine result of actual experiments; of experiments made upon a very large scale, and under circumstances
which render them peculiarly interesting; I cannot help flattering
myself that my readers will find both amusement, and useful information, from the perusal of the following sheets.

As it may perhaps appear extraordinary that a military man
should undertake a work so foreign to his profession, as that of
forming and executing a plan for providing for the Poor, I have

thought it not improper to preface the narrative of my operations, by a short account of the motives which induced me to engage in this undertaking. And in order to throw still more light upon the whole transaction, I shall begin with a few words of myself, of my situation in the country in which I reside, and of the different objects which were had in view in the various public measures in which I have been concerned. This information is necessary in order to form a clear idea of the circumstances under which the operations in question were undertaken, and the different public measures which were adopted at the same time.

Having in the year 1784, with His Majesty's gracious permission, engaged myself in the service of His Most Serene Highness the Elector Palatine, Reigning Duke of Bavaria, I have since been employed by His Electoral Highness in various public services, and particularly in arranging his military affairs, and introducing a new system of order, discipline, and economy among his troops.

In the execution of this commission, ever mindful of that great and important truth, that no political arrangement can be really good, except in so far as it contributes to the general good of society, I have endeavoured in all my operations to unite the interest of the soldier with the interest of civil society, and to render the military force, even in time of peace, subservient to the PUBLIC GOOD.

To facilitate and promote these important objects, to establish a respectable standing military force, which should do the least possible harm to the population, morals, manufactures, and agriculture of the country, it was necessary to make soldiers citizens, and citizens soldiers. To this end the situation of the soldier was made as easy, comfortable, and eligible as possible; his pay was increased, he was comfortably, and even elegantly clothed, and he was allowed every kind of liberty not inconsistent with good order and due subordination; his military exercises were simplified, his instruction rendered short and easy, and all obsolete and useless customs and usages were banished from the service. Great attention was paid to the external appearance of the buildings; and nothing was left undone, that could tend to make the men comfortable in their dwellings. Schools were established in all the regiments, for arithmetic; and into these schools, not only the soldiers and their children, but also the children of the neighbouring citizens and peasants, were admitted gratis, and even school-books, paper[1], pens, and ink, were furnished for them, at the expense of the Sovereign.

Besides these schools of instruction, others, called schools of industry, were established in the regiments, where the soldiers and their children were taught various kinds of work, and from whence they were supplied with raw materials, to work for their own emolument.

As nothing is so certain fatal to morals, and particularly to the morals of the lower class of mankind, as habitual idleness, every possible measure was adopted, that could be devised, to introduce a spirit of industry among the troops. Every encouragement was given to the soldiers to employ their leisure time, when they were off duty, in working for their own emolument; and among other encouragements, the most efficacious of all, that of allowing them full liberty to dispose of the money acquired by their labour in any way they should think proper, without being obliged to give any account of it to any body. They were even furnished with working dresses, (a canvas frock and trousers,) gratis, at their enlisting, and were afterwards permitted to retain their old uniforms for the same purpose; and care was taken, in all cases where they were employed, that they should be well paid.

They commonly received from sixteen to eighteen creutzers[2] a-day for their labour; and with this they had the advantage of being clothed and lodged, and, in many cases, of receiving their full pay of five creutzers, and a pound and a half (1 lb. 13 1/2; oz. Avoirdupois) of bread per day from the Sovereign. When they did their duty in their regiments, by mounting guard regularly according to their tour (which commonly was every fourth day,) and only worked those days they happened to be off guard, in that case, they received their full pay; but when they were excused from regimental duty, and permitted to work every day for their own emolument, their pay (at five creutzers per day,) was stopped, but they were still permitted to receive their bread, and to lodge in the barracks.

In all public works, such as making and repairing highways, — draining marshes, — repairing the banks of rivers, etc. soldiers were employed as labourers; and in all such cases, the greatest care was taken to provide for their comfortable subsistence, and even for their amusement. Good lodgings were prepared for them, and good and wholesome food, at a reasonable price; and the greatest care was taken of them when they happened to fall sick.

Frequently, when considerable numbers of them were at work together, a band of music was ordered to play to them while at work; and on holidays they were permitted, and even encouraged, to make merry, with dancing and other innocent sports and amusements.

To preserve good order and harmony among those who were detached upon these working parties, a certain proportion of officers and non-commissioned officers were always sent with them, and those commonly served as overseers of the works, and as such were paid.

Besides this permission to work for hire in the garrison towns, and upon detached working parties, which was readily granted to all those who desired it, or at least to as many as could possibly be spared from the necessary service of the garrison; every facility and encouragement was given to the soldier who was a native of the country, and who had a family of friends to go to, or private concerns to take care of, to go home on furlough, and to remain absent from his regiment from one annual exercise to the other, that is to say, ten months and a half each year. This arrangement was very advantageous to the agriculture and manufactures, and even to the population of the country, (for the soldiers were allowed to marry,) and served not a little to the establishment of harmony and a friendly intercourse between the soldiers and the peasantry, and to facilitate recruiting.

Another measure which tended much to render the situation of the soldier pleasant and agreeable, and to facilitate the recruiting service, was the rendering the garrisons of the regiments permanent. This measure might not be advisable in a despotic, or odious government; for where the authority of the Sovereign must be supported by the terror of arms, all habits of social intercourse and friendship between the soldiers and the subjects must be dangerous; but in all well-regulated governments, such friendly intercourse is attended with many advantages.

A peasant would more readily consent to his son's engaging himself to serve as a soldier in a regiment permanently stationed in his neighbourhood, than in one at a great distance, or whose destination was uncertain; and when the station of a regiment is permanent, and it receives its recruits from the district of country immediately surrounding its head-quarters, the men who go home on furlough have but a short journey to make, and are easily assembled in case of any emergency; and it was the more necessary to give every facility to the soldiers to go home on furlough in Bavaria, as labourers are so very scarce in that country that the husbandman would not be able without them to cultivate his ground.

The habits of industry and of order which the soldier acquired when in garrison, rendered him so much the more useful as a labourer when on furlough; but not contented with merely furnishing labours for the assistance of the husbandman, I was desirous of making use of the army, as a means of introducing useful improvements into the country.

Though agriculture is carried to the highest perfection in some parts of the Elector's dominions, yet in others, and particularly in Bavaria, it is still much behind-hand. Very few of the new improvements in that art, such as the introduction of new and useful plants — the cultivation of clover and of turnips — the regular succession of crops, etc. have yet found their way into general practice in that country; and even the potatoe, that most useful of all the products of the ground, is scarcely known there.

It was principally with a view to introduce the culture of potatoes in that country that the military gardens were formed. These gardens (of which there is one in every garrison belonging to the Elector's dominion, Dusseldorf and Amberg only excepted[3]) are pieces of ground, in, or adjoining to the garrison towns, which are regularly laid out, and exclusively appropriated to the use of the non-commissioned officers and private soldiers belonging to the regiments in garrison. The ground is regularly divided into districts of regiments, battalions, companies, and corporalities (corporalschafts,) of which last divisions there are four to each company; and the quantity of ground allotted to each corporality is such that each man belonging to it, whether non-commissioned officer or private, has a bed 365 square feet in superficies.

This piece of ground remains his sole property as long as he continues to serve in the regiment, and he is at full liberty to cultivate it in any way, and to dispose of the produce of it in any manner he may think proper. He must however cultivate it, and plant it, and keep it neat and free from weeds; otherwise, if he should be idle, and neglect it, it would be taken from him and given to one of his more industrious comrades.

The divisions of these military gardens are marked by broader and smaller alleys, covered with gravel, and neatly kept; and in order that every one who chooses it, may be a spectator of this interesting scene of industry, all the principal alleys, which are made large for that purpose, are always open as a public walk. The effect which this establishment has already produced in the short time (little more than five years) since it was begun, is very striking, and much greater and more important than I could have expected.

The soldiers, from being the most indolent of mortals, and from having very little knowledge of gardening, or of the produce of a garden, for use, are now becoming industrious and skilful cultivators, and they are grown so fond of vegetables, particularly of potatoes, which they raise in great quantities, that these useful and wholesome productions now constitutes a very essential part of their daily food. And these improvements are also spreading very fast among the farmers and peasants, throughout the whole country. There is hardly a soldier that goes on furlough, or that returns home at the expiration of his time of service, that does not carry with him a few potatoes for planting, and a little collection of garden-seeds; and I have no doubt but in a very few years we shall see potatoes as much cultivated in Bavaria as in other countries; and that the use of vegetables for food will be generally introduced among the common people. I have already had the satisfaction to see little gardens here and there making their appearance, in different parts of the country, and I hope that very soon no farmer's house will be found without one.

To assist the soldiers in the cultivation of their gardens, they are furnished with garden utensils gratis; they are likewise furnished from time to time with a certain quantity of manure, and with an assortment of garden-feeds; but they do not rely solely upon these supplies; those who are industrious collect materials in their barracks, and in the streets, for making manure, and even sometimes purchase it, and they raise in their own gardens most of the garden-seeds they stand in need of. To enable them to avail themselves of their gardens as early in the spring as possible, in supplying their tables with green vegetables, each company is furnished with a hot-bed for raising early plants.

To attach the soldiers more strongly to these their little possessions, by increasing their comfort and convenience in the cultivation and enjoyment of them, a number of little summer-houses, or rather huts, one to each company, have been erected for the purpose of shelter, where they can retire when it rains, or when they are fatigued.

All the officers of the regiments, from the highest to the lowest, are ordered to give the men every assistance in the cultivation of these their gardens; but they are forbidden, upon pain of the severest punishment, to appropriate to themselves any part of the produce of them, or even to receive any part of it in presents.

CHAPTER. I.

Of the prevalence of mendicity in Bavaria at the time when the measures for putting an end to it were adopted.

Among the various measures that occurred to me by which the military establishment of the country might be made subservient to the public good in time of peace, none appeared to be of so much importance as that of employing the army in clearing the country of beggers, thieves and other vagabonds; and in watching over the public tranquillity.

But in order to clear the country of beggers, (the number of whom in Bavaria had become quite intolerable,) it was necessary to adopt general and efficacious measures for maintaining and supporting the Poor. Laws were not wanting to oblige each community in the country to provide for its own Poor; but these laws had been so long neglected, and beggary had become so general, that extraordinary measures, and the most indefatigable exertions, were necessary to put a stop to this evil. The number of itinerant beggars, of both sexes, and all ages, as well foreigners as natives, who strolled about the country in all directions. levying contributions from the industrious inhabitants, stealing and robbing, and leading a life of indolence, and the most shameless debauchery, was quite incredible; and so numerous were the swarms of beggars in all the great towns, and particularly in the capital, so great their impudence, and so persevering their importunity, that it was almost impossible to cross the streets without being attacked, and absolutely forced to satisfy their clamorous demands. And these beggars were in general by no means such as from age or bodily infirmities were unable by their labour to earn their livelihood; but they were for the most part, stout, strong, healthy, sturdy beggars, who, lost to every sense of shame, had embraced the profession from choice, not necessity; and who, not unfrequently, added insolence and threats to their importunity, and extorted that from fear, which they could not procure by their arts of dissimulation.

These beggars not only infested all the streets, public walks, and public places, but they even made a practice of going into private houses, where they never failed to steal whatever fell in their way, if they found the doors open, and nobody at home; and the churches were so full of them that it was quite a nuisance, and a public scandal during the performance of divine service. People at their devotions were continually interrupted by them, and were frequently obliged to satisfy their demands in order to be permitted to finish their prayers in peace and quite.

In short, these detestable vermin swarmed every where, and not only their impudence and clamorous importunity were without any bounds, but they had recourse to the most diabolical arts, and most horrid crimes, in the prosecution of their infamous trade. Young children were stolen from their parents by these wretches, and their eyes put out, or their tender limbs broken and distorted, in order, by exposing them thus maimed, to excite the pity and commiseration of the public; and every species of artifice was made use of to agitate the sensibility, and to extort the contributions of the humane and charitable.

Some of these monsters were so void of all feeling as to expose even their own children, naked, and almost starved, in the streets, in order that, by their cries and unaffected expressions of distress, they might move those who passed by to pity and relieve them; and in order to make them act their part more naturally, they were unmercifully beaten when they came home, by their inhuman parents, if they did not bring with them a certain sum, which they were ordered to collect.

I have frequently seen a poor child of five or six years of age, late at night, in the most inclement season, sitting down almost naked at the corner of a street, and crying most bitterly; if he were asked what was the matter with him, he would answer, "I am cold and hungry, and afraid to go home; my mother told me to bring home twelve creutzers, and I have only been able to beg five. My mother will certainly beat me if I don't carry home twelve creutzers." Who could refuse so small a sum to relieve so much unaffected distress? — But what horrid arts are these, to work upon the feelings of the public, and levy involuntary contributions for the support of idleness and debauchery!

But the evils arising from the prevalence of mendicity did not stop here. The public, worn out and vanquished by the numbers and persevering importunity of the beggars; and frequently disappointed in their hopes of being relieved from their depredations, by the failure of the numberless schemes that were formed and set on foot for that purpose, began at last to consider the case as quite desperate; and to submit patiently to an evil for which they saw no remedy. The consequences of this submission are easy to be conceived; the beggars, encouraged by their success, were attached still more strongly to their infamous profession; and others, allured by their indolent lives, encouraged by their successful frauds, and emboldened by their impunity, joined them. The habit of submission on the part of the public, gave them a sort of right to pursue their depredations;— their growing numbers and their success gave a kind of eclat to their profession; and the habit of begging became so general, that it ceased to be considered as infamous; and was by degrees in a manner interwoven with the internal regulations of society. Herdsmen and shepherds, who attended their flocks by the road-side, were known to derive considerable advantage from the contributions which their situation enabled them to levy from passengers; and I have been assured, that the wages they received from their employers were often regulated accordingly. The children in every country village, and those even of the best farmers, made a constant practice of begging from all strangers who passed; and one hardly ever met a person on foot upon the road, particularly a woman, who did not hold out her hand and ask for charity.

In the great towns, besides the children of the poorer sort, who almost all made a custom of begging, the professional beggars formed a distinct class, or cast, among the inhabitants; and in general a very numerous one. There was even a kind of political connection between the members of this formidable body; and certain general maxims were adopted, and regulations observed, in the warfare they carried on against the public. Each beggar had his particular beat, or district, in the possession of which it was not thought lawful to disturb him; and certain rules were observed in disposing of the districts in case of vacancies by deaths or resignations, promotions or removals. A battle, it is true, frequently decided the contest between the candidates; but when the possession was once obtained, whether by force of arms, or by any other means, the right was ever after considered as indisputable. Alliances by marriage were by no means uncommon in this community; and, strange as it may appear, means were found to procure legal permission from the civil magistrates for the celebration of these nuptials! The children were of course trained up in the profession of their parents; and having the advantage of an early education, were commonly great proficients in their trade.

As there is no very essential difference between depriving a person of his property by stealth, and extorting it from him against his will, by dint of clamorous importunity, or under false pretence of feigned distress and misfortune; so the transition from begging to stealing is not only easy, but perfectly natural. That total insensibility to shame, and all those other qualifications which are necessary in the profession of a beggar, are likewise essential to form an accomplished thief; and both these professions derive very considerable advantages from their union. A beggar who goes about from house to house to ask for alms, has many opportunities to steal, which another would not so easily find; and his profession as a beggar gives him a great facility in disposing of what he steals; for he can always say it was given him in charity. No wonder then that thieving and robbing should be prevalent where beggars are numerous.

That this was the case in Bavaria will not be doubted by those who are informed that in the four years immediately succeeding the introduction of the measures adopted for putting an end to mendicity, and clearing the country of beggars, thieves, robbers, etc. above TEN THOUSAND of these vagabonds, foreigners and natives, were actually arrested and delivered over to the civil magistrates; and that in taking up the beggars in Munich, and providing for those who stood in need of public assistance, no less than 2600 of the one description and the other, were entered upon the lists in one week; though the whole number of the inhabitants of the city of Munich probably does not amount to more than 60,000, even including the suburbs.

These facts are so very extraordinary, that were they not notorious, I should hardly have ventured to mention them, for fear of being suspected of exaggeration; but they are perfectly known in the country, by every body; having been published by authority in the news-papers at the time, with all their various details and specifications, for the information of the public.

What has been said, will, I fancy, be thought quite sufficient to show the necessity of applying a remedy to the evils described; and of introducing order and a spirit of industry among the lower classes of the people. I shall therefore proceed, without any farther preface, to give an account of the measures which were adopted and carried into execution for that purpose.

CHAPTER. II.

Various preparations made for putting an end to mendicity in ba-
varia.
Cantonment of the cavalry in the country towns and villages.
Formation of the committee placed at the head of the institution
  for the poor at Munich.
The funds of that institution.

As soon as it was determined to undertake this great and difficult
work, and the plan of operations was finally settled, various preparations were made for its execution.

The first preliminary step taken, was to canton four regiments of
cavalry in Bavaria and the adjoining provinces, in such a manner
that not only every considerable town was furnished with a detachment, but most of the large villages were occupied; and in every
part of the country small parties of threes, fours, and fives, were so
stationed; at the distance of one, two, and three leagues from each
other; that they could easily perform their daily patroles from one
station to another in the course of the day, without ever being
obliged to stop at a peasant's house, or even at an inn, or ever to
demand forage for their horses, or victuals for themselves, or lodgings, from any person whatever. This arrangement of quarters prevented all disputes between the military and the people of the country. The head-quarters of each regiment, where the commanding
officer of the regiment resided, was established in a central situation
with respect to the extent of country occupied by the regiment;—
each squadron had its commanding officer in the centre of its district,— and the subalterns and non-commissioned officers were so
distributed in the different cantonments, that the privates were
continually under the inspection of their superiors, who had orders
to keep a watchful eye over them;—to visit them in their quarters
very often;—and to preserve the strictest order and discipline
among them.

To command these troops, a general officer was named, who, after visiting every cantonment in the whole country, took up his residence at Munich.

Printed instructions were given to the officer, or non-commissioned officer, who commanded a detached post, or patrole;—regular monthly returns were ordered to be made to the commanding officers of the regiment, by the officers commanding squadrons;— to the commanding general, by the officers commanding regiments;— and by the commanding general, to the council of war, and to the Sovereign.

To prevent disputes between the military and the civil authorities, and, as far as possible, to remove all grounds of jealousy and ill-will between them; as also to preserve peace and harmony between the soldiery and the inhabitants, these troops were strictly ordered and enjoined to behave on all occasions to magistrates and other persons in civil authority with the utmost respect and deference;—to conduct themselves towards the peasants and other inhabitants in the most peaceable and friendly manner;— to retire to their quarters very early in the evening;— and above all, cautiously to avoid disputes and quarrels with the people of the country. They were also ordered to be very diligent and alert in making their daily patroles from one station to another;— to apprehend all thieves and other vagabonds that infested the country, and deliver them over to the civil magistrates;— to apprehend deserters, and conduct them from station to station to their regiments;—to conduct all prisoners from one part of the country to another;—to assist the civil magistrate in the execution of the laws, and in preserving peace and order in the country, in all cases where they should be legally called upon for that purpose;—to perform the duty of messengers in carrying government dispatches and orders, civil as well as military, in cases of emergency;— and to bring accounts to the capital, by express, of every extraordinary event of importance that happens in the country;—to guard the frontiers, and assist the officers of the revenue in preventing smuggling;—to have a watchful eye over all soldiers on furlough in the country, and when guilty of excesses, to apprehend them and transport them to their regiments;—to assist the inhabitants in case of fire, and particularly to guard their effects, and prevent their being lost of stolen, in the confusion which commonly takes place on those occasions; — to pursue and apprehend all thieves, robbers, murderers, and other malefactors; — and in general, to lend their assistance on all occasions where they could be useful in maintaining peace, order, and tranquillity in the country.

As the Sovereign had an undoubted right to quarter his troops upon the inhabitants when they were employed for the police and defence of the country, they were on this occasion called upon to provide quarters for the men distributed in these cantonments; but in order to make this burden as light as possible to the inhabitants, they were only called upon to provide quarters for the non-commissioned officers and privates; and instead of being obliged to take THESE into their houses, and to furnish them with victuals and lodgings, as had formerly been the practice, (and which was certainly a great hardship,) a small house or barrack for the men, with stabling adjoining to it for the horses, was built, or proper lodgings were hired by the civil magistrate, in each of these military stations, and the expense was levied upon the inhabitants at large. The forage for the horses was provided by the regiments, or by contractors employed for that purpose; and the men, being furnished with a certain allowance of fire-wood, and the necessary articles of kitchen furniture, were made to provide for their own subsistence, by purchasing their provisions at the markets, and cooking their victuals in their own quarters.

The officers provided their own lodgings and stabling, being allowed a certain sum for that purpose in addition to their ordinary pay.

The whole of the additional expence to the military chest, for the establishment and support of these cantonments, amounted to a mere trifle; and the burden upon the people, which attended the furnishing of quarters for the non-commissioned officers and privates, was very inconsiderable, and bore no proportion to the advantages derived from the protection and security to their persons and properties afforded by these troops[4].

Not only this cantonment of the cavalry was carried into execution as a preliminary measure to the taking up of the beggars in the capital, but many other preparatives were also made for that undertaking.

As considerable sums were necessary for the support of such of the poor as, from age or other bodily infirmities, were unable by their industry to provide for their own subsistence; and as there were no public funds any way adequate to such an expence, which could be applied to this use, the success of the measure depended entirely upon the voluntary subscriptions of the inhabitants; and in order to induce these to subscribe liberally, it was necessary to secure their approbation of the plan, and their confidence in those who were chosen to carry it into execution. And as the number of beggars was so great in Munich, and their importunity so very troublesome, there could have been no doubt but any sensible plan for remedying this evil would have been gladly received by the public; but they had been so often disappointed by fruitless attempts from time to time made for that purpose, that they began to think the enterprize quite impossible, and to consider every proposal for providing for the poor, and preventing mendicity, as a mere job.

Aware of this, I took my measures accordingly. To convince the public that the scheme was feasible, I determined first, by a great exertion, to carry it into complete execution, and THEN to ask them to support it. And to secure their confidence in those employed in the management of it, persons of the highest rank, and most respected character were chosen to superintend and direct the affairs of the institution; and every measure was taken that could be devised to prevent abuses.

Two principle objects were to be attended to, in making these arrangements; the first was to furnish suitable employment to such of the poor as were able to work; and the second, to provide the necessary assistance for those who, from age, sickness, or other bodily infirmities, were unable by their industry to provide for themselves. A general system of police was likewise necessary among this class of miserable beings; as well as measures for reclaiming them, and making them useful subjects. The police of the poor, as also the distribution of alms, and all the economical details of the institution, were put under the direction of a committee, composed of the president of the council of war, — the president of the council of supreme regency, — the president of the ecclesiastical council, — and the president of the chamber of finances; and to assist them in this work, each of the above-mentioned presidents was accompanied by one counsellor of his respective department, at his own choice; who was present at all the meetings of the committee, and who performed the more laborious parts of the business. This committee, which was called The Armen Instituts Deputation, had convenient apartments fitted up for its meetings; a secretary, clerk, and accountant, were appointed to it; and the ordinary guards of the police were put under its immediate direction.

Neither the presidents nor the counsellors belonging to this committee received any pay or emolument whatever for this service, but took upon themselves this trouble merely from motives of humanity, and a generous desire to promote the public good; and even the secretary, and other inferior officers employed in this business, received their pay immediately from the Treasury; or from some other department; and not from the funds destined for the relief of the poor: and in order most effectually remove all suspicion with respect to the management of this business, and the faithful application of the money destined for the poor, instead of appointing a Treasurer to the committee, a public banker of the town, a most respectable citizen[5], was named to receive and pay all monies belonging to the institution, upon the written orders of the committee; and exact and detailed accounts of all monies received and expended were ordered to be printed every three months, and distributed gratis among the inhabitants.

In order that every citizen might have it in his power to assure himself that the accounts were exact, and that the sums expended were bona fide given to the poor in alms, the money was publicly distributed every Saturday in the town-hall, in the presence of a number of deputies chosen from among the citizens themselves; and an alphabetical list of the poor who received alms;—in which was mentioned the weekly sum each person received;—and the place of his or her abode, was hung up in the hall for public inspection.

But this was not all. In order to fix the confidence of the public upon the most firm and immoveable basis, and to engage their good will and cheerful assistance in support of the measures adopted, the citizens were invited to take an active and honourable part in the execution of the plan, and in the direction of its most interesting details.

The town of Munich, which contains about 60,000 inhabitants, had been formerly divided into four quarters. Each of these was now subdivided into four districts, making in all sixteen districts; and all the dwelling-houses, from the palace of the sovereign to the meanest hovel, were regularly numbered, and inscribed in printed lists provided for that purpose. For the inspection of the poor in each district, a respectable citizen was chosen, who was called the commissary of the district, (abtheilungs commissaire,) and for his assistance, a priest; a physician; a surgeon; and an apothecary; all of whom, including the commissary, undertook this service without fee or reward, from mere motives of humanity and true patriotism. The apothecary was simply reimbursed the original cost of the medicines he furnished.

To give more weight and dignity to the office of commissary of a district, one of these commissaries, in rotation, was called to assist at the meetings of the supreme committee; and all applications for alms were submitted to the commissaries for their opinion; or, more properly, all such applications went through them to the committee. They were likewise particularly charged with the inspection and police of the poor in their several districts.

When a person already upon the poor list, or any other, in distress, stood in need of assistance, he applied to the commissary of his district, who, after visiting him, and enquiring into such the circumstances of his case, afforded him such immediate assistance as was absolutely necessary; or otherwise, if the case was such as to admit of the delay, he recommended him to the attention of the committee, and waited for their orders. If the poor person was sick, or wounded, he was carried to some hospital; or the physician, or surgeon of the district was sent for, and a nurse provided to take care of him in his lodgings, If he grew worse, and appeared to draw near his end, the priest was sent for, to afford him such spiritual assistance as he might require; and if he died, he was decently buried. After his death, the commissary assisted at the inventory which was taken of his effects, a copy of which inventory was delivered over to the committee. These effects were afterwards sold;—and after deducting the amount of the different sums received in alms from the institution by the deceased during his lifetime, and the amount of the expenses of his illness and funeral, the remainder, if any, was delivered over to his lawful heirs; but when these effects were insufficient for those purposes; or when no effects were to be found, the surplus in the one case, and the whole of these expences in the other, was borne by the funds of the institution.

These funds were derived from the following sources, viz.

First, from stated monthly allowances, from the sovereign out of his private purse,—from the states,—and from the treasury, or chamber of finances.

Secondly, and principally, from the voluntary subscription of the inhabitants.

Thirdly, from legacies left to the institution, and

Fourthly, from several small revenues arising from certain tolls, fines, etc. which were appropriated to that use[6].

Several other, and some of them very considerable public funds, originally designed by their founders for the relief of the poor, might have been taken and appropriated to this purpose; but, as some of these foundations had been misapplied, and others nearly ruined by bad management, it would have been a very disagreeable task to wrest them out of the hands of those who had the administration of them; and I therefore judged it most prudent not to meddle with them, avoiding, by that means, a great deal of opposition to the execution of my plan.

CHAPTER. III.

Preparations made for giving employment to the poor.
Difficulties attending that undertaking.
The measures adopted completely successful.
The poor reclaimed to habits of useful industry.
Description of the house of industry at Munich.

But before I proceed to give a more particular account of the funds of this institution, and of the application of them, it will be necessary to mention the preparations which where made for furnishing employment to the poor, and the means which were used for reclaiming them from their vicious habits, and rendering them industrious and useful subjects. And this was certainly the most difficult, as well as the most curious and interesting part of the undertaking. To trust raw materials in the hands of common beggars, certainly required great caution and management; —but to produce so total and radical a change in the morals, manners, and customs of this debauched and abandoned race, as was necessary to render them orderly and useful members of society, will naturally be considered as an arduous, if not impossible, enterprize. In this I succeeded; —for the proof of this fact I appeal to the flourishing state of the different manufactories in which these poor people are now employed,—to their orderly and peaceable demeanour—to their cheerfulness—to their industry,— to the desire to excel, which manifests itself among them upon all occasions,—and to the very air of their countenances. Strangers, who go to see this institution, (and there are very few who pass through Munich who do not take that trouble,) cannot sufficiently express their surprise at the air of happiness and contentment which reigns throughout every part of this extensive establishment, and can hardly be persuaded, that among those they see so cheerfully engaged in that interesting scene of industry, by far the greater part were, five years ago, the most miserable and most worthless of beings,—common beggars in the streets.

An account of the means employed in bringing about this change cannot fail to be interesting to every benevolent mind; and this is what has encouraged me to lay these details before the public.

By far the greater number of the poor people to be taken care of were not only common beggars, but had been up from their very infancy in that profession; and were so attached to their indolent and dissolute way of living, as to prefer it to all other situations. They were not only unacquainted with all kinds of work, but had the most insuperable aversion to honest labour; and had been so long familiarized with every crime, that they had become perfectly callous to all sense of shame and remorse.

With persons of this description, it is easy to be conceived that precepts;—admonitions;—and punishments, would be of little or no avail. But where precepts fail, HABITS may sometimes be successful.

To make vicious and abandoned people happy, it has generally been supposed necessary, FIRST to make them virtuous. But why not reverse this order? Why not make them first HAPPY, and then virtuous? If happiness and virtue be INSEPARABLE the end will be as certainly obtained by the one method as by the other; and it is most undoubtedly much easier to contribute to the happiness and comfort of persons in a state of poverty and misery, than, by admonitions and punishments, to reform their morals.

Deeply struck with the importance of this truth, all my measures were taken accordingly. Every thing was done that could be devised to make the poor people I had to deal with comfortable and happy in their new situation; and my hopes, that a habit of enjoying the real comforts and conveniences which were provided for them, would in time, soften their hearts;—open their eyes;—and render them grateful and docile, were not disappointed.

The pleasure I have had in the success of this experiment is much easier to be conceived than described. Would God that my success might encourage others to follow my example! If it were generally known how little trouble, and how little expence, are required to do much good, the heart-felt satisfaction which arises from relieving the wants, and promoting the happiness of our fellow-creatures, is so great, that I am persuaded, acts of the most essential charity would be much more frequent, and the mass of misery among mankind would consequently be much lessened.

Having taken my resolution to make the COMFORT of the poor people, who were to be provided for, the primary object of my attention, I considered what circumstance in life, after the necessaries, food and raiment, contributes most to comfort, and I found it to be CLEANLINESS. And so very extensive is the influence of cleanliness, that it reaches even to the brute creation.

With what care and attention do the feathered race wash themselves and put their plumage in order; and how perfectly neat, clean and elegant do they ever appear! Among the beasts of the field we find that those which are the most cleanly are generally the most gay and cheerful; or are distinguished by a certain air of tranquillity and contentment; and singing birds are always remarkable for the neatness of their plumage. And so great is the effect of cleanliness upon man, that it extends even to his moral character. Virtue never dwelt long with filth and nastiness; nor do I believe there ever was a person SCRUPULOUSLY ATTENTIVE TO CLEANLINESS who was a consummate villain[7].

Order and disorder — peace and war — health and sickness, cannot exist together; but COMFORT and CONTENTMENT the inseparable companions of HAPPINESS and VIRTUE, can only arise from order, peace, and health.

Brute animals are evidently taught cleanliness by instinct; and can there be a stronger proof of its being essentially necessary to their well-being and happiness? — But if cleanliness is necessary to the happiness of brutes, how much more so must it be to the happiness of the human race?

The good effects of cleanliness, or rather the bad effects of filth and nastiness, may, I think, be very satisfactorily accounted for. Our bodies are continually at war with whatever offends them, and every thing offends them that adheres to them, and irritates them, — and through by long habit we may be so accustomed to support a physical ill, as to become almost insensible to it, yet it never leaves the mind perfectly at peace. There always remains a certain uneasiness, and discontent; — an indecision, and an aversion from all serious application, which shows evidently that the mind is not at rest.

Those who from being afflicted with long and painful disease, suddenly acquire health, are best able to judge of the force of this reasoning. It is by the delightful sensation they feel, at being relieved from pain and uneasiness, that they learn to know the full extent of their former misery; and the human heart is never so effectually softened, and so well prepared and disposed to receive virtuous impressions, as upon such occasions.

It was with a view to bring the minds of the poor and unfortunate people I had to deal with to this state, that I took so much pains to make them comfortable in their new situation. The state in which they had been used to live was certainly most wretched and deplorable; but they had been so long accustomed to it, that they were grown insensible to their own misery. It was therefore necessary, in order to awaken their attention, to make the contrast between their former situation, and that which was prepared for them, as striking as possible. To this end, every thing was done that could be devised to make them REALLY COMFORTABLE.

Most of them had been used to living in the most miserable hovels, in the midst of vermin, and every kind of filthiness; or to sleep in the streets, and under the hedges, half naked, and exposed to all the inclemencies of the seasons. A large and commodious building, fitted up in the neatest and most comfortable manner, was now provided for their reception. In this agreeable retreat they found spacious and elegant apartments, kept with the most scrupulous neatness; well warmed in winter; well lighted; a good warm dinner every day, gratis; cooked and served up with all possible attention to order and cleanliness;— materials and utensils for those who required instruction;—the most generous pay, IN MONEY, for all the labour performed; and the kindest usage from every person, from the highest to the lowest, belonging to the establishment. Here, in this asylum for the indigent and unfortunate, no ill usage;— no harsh language, is permitted. During five years that the establishment has existed, not a blow has been given to any one; not even to a child by his instructor.

As the rules and regulations for the preservation of order are few, and easy to be observed, the instances of their being transgressed are rare; and as all the labour performed, is paid for by the piece;

and not by the day; and is well paid; and as those who gain the most by their work in the course of the week, receive proportional rewards on the Saturday evening; these are most effectual encouragements to industry.

But before I proceed to give an account of the internal economy of this establishment, it will be necessary to describe the building which was appropriated to this use; and the other local circumstances, necessary to be known, in order to have a clear idea of the subject.

This building, which is very extensive, is pleasantly situated in the Au, one of the suburbs of the city of Munich. It had formerly been a manufactory, but for many years had been deserted and falling to ruins. It was now completely repaired, and in part rebuilt. A large kitchen, with a large eating-room adjoining it, and a commodious bake-house, were added to the buildings; and such other mechanics as were constantly wanted in the manufactory for making and repairing the machinery were established, and furnished with tools. Large halls were fitted up for spinners of hemp;—for spinners of flax;—for spinners of cotton;—for spinners of wool;—and for spinners of worsted; and adjoining to each hall a small room was fitted up for a clerk or inspector of the hall, (spin-schreiber). This room, which was at the same time a store-room, and counting-house, and a large window opening into the hall, from whence the spinners were supplied with raw materials;—where they delivered their yarn when spun;—and from whence they received an order upon the cashier, signed by the clerk, for the amount of their labour.

Halls were likewise fitted up for weavers of woollens;— for weavers of serges and shalloons;—for linen weavers;— for weavers of cotton goods, and for stocking weavers;— cloth shearers;—dryers;—sadlers;—wool-combers;—knitters;— sempstresses, etc. Magazines were fitted up as well for finished manufactures, as for raw materials, and rooms for counting-houses, —store-rooms for the kitchen and bake-house,—and dwelling-rooms for the inspectors and other officers who were lodged in the house.

A very spacious hall, 110 feet long, 37 feet wide, and 22 feet high, with many windows on both sides, was fitted up as a drying-room; and in this hall tenters were placed for stretching out and drying eight pieces of cloth at once. The hall was so contrived as to serve for the dyer and for the clothier at the same time.

A fulling-mill was established upon a stream of water which runs by one side of the court round which the building is erected; and adjoining to the fulling-mill, is the dyers-shop; and the wash-house.

This whole edifice, which is very extensive, was fitted up, as has already been observed, in the neatest manner possible. In doing this, even the external appearance of the building was attended to. It was handsomely painted; without, as well as within; and pains were taken to give it an air of ELEGANCE, as well as of neatness and cleanliness. A large court in the middle of the building was levelled, and covered with gravel; and the approach to it from every side was made easy and commodious. Over the principal door, or rather gate, which fronts the street, is an inscription, denoting the use to which the building is appropriated; and the passage leading into the court, there is written in large letters of gold upon a black ground— "NO ALMS WILL BE RECEIVED HERE."

Upon coming into the court you see inscriptions over all the doors upon the ground floor, leading to the different parts of the building. These inscriptions, which are all in letters of gold upon a black ground, denote the particular uses to which the different apartments are destined.

This building having been got ready, and a sufficient number of spinning-wheels, looms, and other utensils made use of in the most common manufactures being provided; together with a sufficient stock of raw materials, I proceeded to carry my plan into execution in the manner which will be related in the following Chapter.

CHAPTER. IV.

An account of the taking up of the beggars at Munich.
The inhabitants are called upon for their assistance.
General subscription for the relief and support of the poor.
All other public and private collections for the poor abolished.

New-Year's-Day having, from time immemorial, been considered in Bavaria as a day peculiarly set apart for giving alms; and the beggars never failing to be all out upon that occasion; I chose that moment as being the most favourable for beginning my operations. Early in the morning of the first of January 1790, the officers and non-commissioned officers of the three regiments of infantry in garrison, were stationed in the different streets, where they were directed to wait for further orders.

Having, in the mean time, assembled, at my lodgings, the field-officers, and all the chief magistrates of the town, I made them acquainted with my intention to proceed that very morning to the execution of a plan I had formed for taking up the beggars, and providing for the poor; and asked their immediate assistance.

To show the public that it was not my wish to carry this measure into execution by military force alone, (which might have rendered the measure odious,) but that I was disposed to show all becoming deference to the civil authority, I begged the magistrates to accompany me, and the field-officers of the garrison, in the execution of the first and most difficult part of the undertaking, that of arresting the beggars. This they most readily consented to, and we immediately sallied out into the street, myself accompanied by the chief magistrate of the town, and each of the field-officers by an inferior magistrate.

We were hardly got into the street when we were accosted by a beggar, who asked us for alms. I went up to him, and laying my hand gently upon his shoulder, told him, that from thenceforwards begging would not be permitted in Munich;—that if he really stood in need of assistance, (which would immediately be enquired into,) the necessary assistance should certainly be given him, but that begging was forbidden; and if he was detected in it again he would be severely punished. I then delivered him over to an orderly serjeant who was following me, with directions to conduct him to the Town-hall, and deliver him into the hands of those he should find there to receive him; and then turning to the officers and magistrates who accompanied me, I begged they would take notice, that I had myself, WITH MY OWN HANDS, arrested the first beggar we had met; and I requested them not only to follow my example themselves, by arresting all the beggars they should meet with, but that they would also endeavour to persuade others, and particularly the officers, non-commissioned officers, and soldiers of the garrison, that it was by no means derogatory to their character as soldiers, or in anywise disgraceful to them, to assist in so USEFUL and LAUDABLE an undertaking. These gentlemen having cheerfully and unanimously promised to do their utmost to second me in this business, dispersed into the different parts of the town, and with the assistance of the military, which they found every where waiting for orders, the town was so thoroughly cleared of beggars IN LESS THAN AN HOUR, that not one was to be found in the streets.

Those who were arrested were conducted to the Town-hall, where their names were inscribed in printed lists provided for that purpose, and they were then dismissed to their own lodgings, with directions to repair the next day to the newly erected "Military Work-house" in the Au; where they would find comfortable warm rooms;—a good warm dinner every day; and work for all those who were in a condition to labour. They were likewise told that a commission should immediately be appointed to enquire into their circumstances, and to grant them such regular weekly allowances of money, in alms, as they should stand in need of; which was accordingly done.

Orders were then issued to all the military guards in the different parts of the town, to send out patroles frequently into the streets in their neighbourhood, to arrest all the beggars they should meet with, and a reward was offered for each beggar they should arrest and deliver over to the civil magistrate. The guard of the police was likewise directed to be vigilant; and the inhabitants at large, of all ranks and denominations, were earnestly called upon to assist in completing a work of so much public utility, and which had been so happily begun[8]. In an address to the public, which was printed and distributed gratis among the inhabitants, the fatal consequences arising from the prevalence of mendicity were described in the most lively and affecting colours,—and the manner pointed out in which they could most effectually assist in putting an end to an evil equally disgraceful and prejudicial to society.

As this address, (which was written with great sprit, by a man well known in the literary world, Professor Babo,) gives a very striking and a very just picture of the character, manners, and customs, of the hords of idle and dissolute vagabonds which infested Munich at the time the measure in question was adopted, and of the various artifices they made use of in carrying on their depredations; I have thought it might not be improper to annex it, at full length, in the Appendix, No. I.

This address, which was presented to all the heads of families in the city, and to many by myself, having gone round to the doors of most of the principal citizens for that purpose, was accompanied by printed lists, in which the inhabitants were requested to set down their names;—places of abode;—and the sums they chose to contribute monthly, for the support of the establishment. These lists, (translations of which are also inserted in the Appendix, No. II.) were delivered to the heads of families, with duplicates, to the end that one copy being sent in to the committee, the other might remain with the master of the family.

These subscriptions being PERFECTLY VOLUNTARY, might be augmented or diminished at pleasure. When any person chose to alter his subscription, he sent to the public office for two blank subscription lists, and filling them up anew, with such alterations as he thought proper to make, he took up his old list at the office, and deposited the new one in its stead.

The subscription lists being all collected, they were sorted, and regularly entered according to the numbers of the houses of the subscribers, in sixteen general lists[9], answering to the sixteen subdivisions or districts of the city; and a copy of the general list of each district was given to the commissary of the district.

These copies, which were properly authenticated, served for the direction of the commissary in collecting the subscriptions in his district, which was done regularly the last Sunday morning of every month.

The amount of the collection was immediately delivered by the commissary into the hands of the banker of the institution, for which he received two receipts from the banker; one of which he transmitted to the committee, with his report of the collection, which he was directed to send in as soon as the collection was made.

As there were some persons who, from modesty, or other motives, did not choose to have it known publicly how much they gave in alms to the poor, and on that account were not willing to have put down to their names upon the list of the subscribers, the whole sum they were desirous of appropriating to that purpose; to accommodate matters to the peculiar delicacy of their feelings, the following arrangement was made, and carried into execution with great success.

Those who were desirous of contributing privately to the relief of the poor, were notified by an advertisement published in the newspapers, that they might send to the banker of the institution any sums for that purpose they might think proper, under any feigned name, or under any motto or other device; and that not only a receipt would be given to the bearer, for the amount, without and questions being asked him, but, for greater security, a public acknowledgement of the receipt of the sum would be published by the banker, with a mention of the feigned name of device under which it came, IN THE NEXT MUNICH GAZETTE.

To accommodate those who might be disposed to give trifling sums occasionally, for the relief of the poor, and who did not choose to go, or to send to the banker, fixed poor-boxes were placed in all the churches, and most of the inns; coffee-houses; and other places of public resort; but nobody was ever called upon to put any thing into these boxes, nor was any poor's-box carried round, or any private collection or alms-gathering permitted to be made upon any occasion, or under any pretence whatever.

When the inhabitants had subscribed liberally to the support of the institution, it was but just to secure them from all further importunity in behalf of the poor. This was promised, and it was most effectually done; though not without some difficulty, and a very considerable expence to the establishment.

The poor students in the Latin German schools;—the sisters of the religious order of charity;—the directors of the hospital of lepers;—and some other public establishments, had been so long in the habit of making collections, by going round among the inhabitants from house to house at stated periods, asking alms, that they had acquired a sort of right to levy those periodical contributions, of which it was not thought prudent to dispossess them without giving them an equivalent. And in order that this equivalent might not appear to be taken from the sums subscribed by the inhabitants for the support of the poor, it was paid out of the monthly allowance which the institution received from the chamber of finances, or public treasury of the state.

Besides these periodical collections, there were others, still more troublesome to the inhabitants, from which it was necessary to free them; and some of these last were even sanctioned by legal authority. It is the custom in Germany for apprentices in most of the mechanical trades, as soon as they have finished their apprenticeships with their masters, to travel, during three or four years, in the neighbouring countries and provinces, to perfect themselves in their professions by working as journeymen wherever they can find employment. When one of those itinerant journeymen-tradesmen comes into a town, and cannot find employment in it, he is considered AS HAVING A RIGHT to beg the assistance of the inhabitants, and particularly of those of the trade he professes, to enable him to go to the next town; and this assistance it was not thought just to refuse. This custom was not only very troublesome to the inhabitants, but gave rife to innumerable abuses. Great numbers of idle vagabonds were continually strolling about the country under the name of travelling journeymen-tradesmen; and though any person, who presented himself as such in any strange place was obliged to produce (for his legitimation) a certificate from his last master, in whose service he had been employed, yet such certificates were so easily counterfeited, or obtained by fraud, that little reliance could be placed in them.

To remedy all these evils, the following arrangement was made: those travelling journeymen-tradesmen who arrive at Munich, and do not find employment, are obliged to quit the town immediately, or to repair to the military work-house, where they are either furnished with work, or a small sum is given them to enable them to pursue their journey farther.

Another arrangement by which the inhabitants have been relieved from much importunity, and by which a stop has been put to many abuses, is the new regulation respecting those who suffer by fire; such sufferers commonly obtain from government special permission to make collections of charitable donations among the inhabitants in certain districts, during a limited time. Instead of the permission to make collections in the city of Munich, the sufferers now receive certain sums from the funds of the institution for the poor. By this arrangement, not only the inhabitants are relieved from the importunity which always attends public collections of alms, but the sufferers save a great deal of time, which they formerly spent in going about from house to house; and the sale of these permissions to undertakers, and many other abuses, but too frequent before this arrangement took place, are now prevented.

The detailed account published in the Appendix, No. III. of the receipts and expenditures of the institution during five years, will show the amount of the expense incurred in relieving the inhabitants from the various periodical and other collections before mentioned.

But not to lose sight too long of the most interesting object of this establishment, we must follow the people who were arrested in the streets, to the asylum which was prepared for them, but which no doubt appeared to them at first a most odious prison.

CHAPTER. V.

The different kinds of employment given to the beggars upon their
  being assembled in the house of industry.
Their great awkwardness at first.
Their docility, and their progress in useful industry.
The manner in which they were treated.
The manner in which they were fed.
The Precautions used to prevent Abuses in the Public Kitchen from
  which they were fed.

As by far the greater part of these poor creatures were totally unacquainted with every kind of useful labour, it was necessary to give them such work, at first, as was very easy to be performed, and in which the raw materials were of little value; and then, by degrees, as they became more adroit, to employ them in manufacturing more valuable articles.

As hemp is a very cheap commodity, and as the spinning of hemp is easily learned, particularly when it is designed for very coarse and ordinary manufactures, 15,000 pounds of that article were purchased in the palatinate, and transported to Munich; and several hundred spinning wheels, proper for spinning it, were provided; and several good spinners, as instructors, were engaged, and in readiness, when this house of industry was opened for the reception of the poor.

Flax and wool were likewise provided, and some few good spinners of those articles were engaged as instructors; but by far the greater number of the poor began with spinning of hemp; and so great was their awkwardness at first, that they absolutely ruined almost all the raw materials that were put into their hands. By an exact calculation of profit and loss, it was found that the manufactory actually lost more than 3000 florins upon the articles of hemp and flax, during the first three months; but we were not discouraged by these unfavourable beginnings; they were indeed easy to be foreseen, considering the sort of people we had to deal with, and how necessary it was to pay them at a very high rate for the little work they were able to perform, in order to persevere with cheerfulness in acquiring more skill and address in their labour. If the establishment was supported at some little expence in the beginning, it afterwards richly repaid these advances, as will be seen in the sequel of this account.

As the clothing of the army was the market upon which I principally depended, in disposing of the manufactures which should be made in the house, the woollen manufactory was an object most necessary to be attended to, and from which I expected to derive most advantage to the establishment; but still it was necessary to begin with the manufacture of hemp and flax, not only because those articles are less valuable than wool, and the loss arising from their being spoiled by the awkwardness of beginners is of less consequence, but also for another reason, which appears to me to be of so much more importance as to require a particular explanation.

It was hinted above that it was found necessary, in order to encourage beginners in these industrious pursuits, to pay them at a very high rate for the little work they were able to perform; but every body knows that no manufacture can possibly subsist long, where exorbitant prices are paid for labour; and it is easy to conceive what discontent and disgust would be occasioned among the workmen upon lowering the prices which had for a length of time been given for labour, By employing the poor people in question at first in the manufactures of hemp and flax, manufactures which were not intended to be carried on to any extent, it was easy afterwards, when they had acquired a certain degree of address in their work, to take them from these manufactures, and put them to spinning of wool, worsted, or cotton; care having been taken to fix the price of labour in these last-mentioned manufactures at a reasonable rate.

The dropping the manufacture of any particular article altogether, or pursuing it less extensively, could produce no bad effect upon the general establishment; but the lowering of the price of labour, in any instance, could not fail to produce many.

It is necessary, in an undertaking like this, cautiously to avoid every thing that could produce discouragement and discontent among those upon whose industry alone success must depend.

It is easy to conceive that so great a number of unfortunate beings, of all ages and sexes, taken as it were out of their very element,
and placed in a situation so perfectly new to them, could not fail to
be productive of very interesting situations. Would to God I were
able to do justice to this subject! but no language can describe the
affecting scenes to which I was a witness upon this occasion.

The exquisite delight which a sensible mind must feel, upon seeing many hundreds of wretched being awaking from a state of misery and inactivity, as from a dream; and applying themselves with
cheerfulness to the employments of useful industry;—upon seeing
the first dawn of placid content break upon a countenance covered
with habitual gloom, and furrowed and distorted by misery;— this
is easier to be conceived than described.

During the first three or four days that these poor people were assembled, it was not possible entirely to prevent confusion: there was
nothing like mutinous resistance among them; but their situation
was so new to them, and they were so very awkward in it, that it
was difficult to bring them into any tolerable order. At length, however, by distributing them in the different halls, and assigning to
each his particular place, (the places being all distinguished by
numbers,) they were brought into such order as to enable the inspectors, and instructors, to begin their operations.

Those who understood any kind of work, were placed in the
apartments where the work they understood was carried on; and
the others, being classed according to their sexes, and as much as
possible according to their ages, were placed under the immediate
care of the different instructors. By much the larger number were
put to spinning of hemp;—others, and particularly the young children from four to seven years of age, were taught to knit, and to
sew; and the most awkward among the men, and particularly the
old, the lame, and the infirm, were put to the carding of wool. Old
women, whose sight was too weak to spin, or whose hands trem-
bled with palsy, were made to spool yarn for the weavers; and
young children, who were too weak to labour, were placed upon
seats erected for that purpose round the rooms where other children
worked.

As it was winter, fires were kept in every part of the building, from morning till night; and all the rooms were lighted up till nine o'clock in the evening. Every room and every stair-case was neatly swept and cleaned twice a day; one early in the morning, before the people were assembled, and once while they were at dinner.—Care was taken, by placing ventilators, and occasionally opening the windows, to keep the air of the rooms perfectly sweet, and free from all disagreeable smells; and the rooms themselves were not only neatly white-washed and fitted up, and arranged in every respect with elegance, but care was taken to clean the windows very often;—to clean the courtyard every day;— and even to clear away the rubbish from the street in front of the building, to a considerable distance on every side.

Those who frequented this establishment were expected to arrive at the fixed hour in the morning, which hour varied according to the season of the year; if they came too late, they were gently reprimanded; and if they persisted in being tardy, without being able to give a sufficient excuse for not coming sooner, they were punished by being deprived of their dinner, which otherwise they received every day gratis.

At the hour of dinner, a large bell was rung in the court, when those at work in the different parts of the building repaired to the dining-hall; where they found a wholesome and nourishing repast; consisting of about A POUND AND A QUARTER, Avoirdupois weight, of a very rich soup of peas and barley, mixed with cuttings of fine white bread; and a piece of excellent rye bread, weighing SEVERN OUNCES; which last they commonly put in their pockets, and carried home for their supper. Children were allowed the same portion as grown persons; and a mother, who had one or more young children, was allowed a portion for each of them.

Those who, from sickness, or other bodily infirmities, were not able to come to the work-house;—as also those who, on account of young children they had to nurse, or sick persons to take care of, found it more convenient to work at their own lodgings, (and of these there were many,) were not on that account deprived of their dinners. Upon representing their cases to the committee, tickets were granted them, upon which they were authorized to receive from the public kitchen, daily, the number of portions specified in the ticket; and these they might send for by a child, or by any other person they thought proper to employ; it was necessary, however that the ticket should always be produced, otherwise the portions were not delivered. This precaution was necessary, to prevent abuses on the part of the poor.

Many other precautions were taken to prevent frauds on the part of those employed in the kitchen, and in the various other offices and departments concerned in feeding the poor.

The bread-corn, peas, barley, etc. were purchased in the public market in large quantities, and at times when those articles were to be had at reasonable prices, and were laid up in store-rooms provided for that purpose, under the care of the store-keeper of the Military Work-house.

The baker received his flour by weight from the store-keeper, and in return delivered a certain fixed quantity of bread. Each loaf, when well baked, and afterwards dried, during four days, in a bread-room through which the air had a free passage, weighed two pounds ten ounces Avoirdupois. Such a loaf was divided into six portions; and large baskets filled with these pieces being placed in the passage leading to the dining-hall, the portions were delivered out to the poor as they passed to go into the hall, each person who passed giving a medal of tin to the person who gave him the bread, in return for each portion received. These medals, which were given out to the poor each day in the halls where they worked, by the steward, or by the inspectors of the hall, served to prevent frauds in the distribution of the bread; the person who distributed it being obliged to produce them as vouchers of the quantity given out each day.

Those who had received these portions of bread, held them up in their hands upon their coming into the dining-hall, as a sign that they had a right to seat themselves at the tables; and as many portions of bread as they produced, so many portions of soup they were entitled to receive; and those portions which they did not eat they were allowed to carry away; so that the delivery of bread was a check upon the delivery of soup, and VICE VERSA.

The kitchen was fitted up with all possible attention, as well to conveniences, as to the economy of fuel. This will readily be believed by those who are informed, that the whole work of the kitchen is performed, with great ease, by three cook-maids; and that the daily expence for fire-wood amounts to no more than twelve creutzers, or FOUR-PENCE HALFPENNY sterling, when dinner is provided for 1000 people. The number of persons who are fed DAILY from this kitchen is, at a medium, in summer, about ONE THOUSAND, (rather more than less,) and in winter, about 1200. Frequently, however, there have been more than 1500 at table. As a particular account of this kitchen, with drawings; together with an account of a number of new and very interesting experiments relative to the economy of fuel, will be annexed to this work, I shall add nothing more now upon the subject; except it be the certificate, which may be seen in the Appendix, No. IV; which I have thought prudent to publish, in order to prevent my being suspected of exaggeration in displaying the advantages of my economical arrangements.

The assertion, that a warm dinner may be cooked for 1000 persons, at the trifling expence of four-pence halfpenny for fuel; and that, too, where the cord, five feet eight inches and nine-tenths long, five feet eight inches and nine-tenths high, and five feet three inches and two-tenths wide, English measure, of pine-wood, of the most indifferent quality, costs above seven shillings; and where the cord of hard wood, such as beech and oak, of equal dimensions, costs more than twice that sum, may appear incredible; yet I will venture to assert, and I hereby pledge myself with the public to prove, that in the kitchen of the Military Academy at Munich, and especially in a kitchen lately built under my direction at Verona, in the Hospital of la Pieta, I have carried the economy of fuel still further.

To prevent frauds in the kitchen of the institution for the poor at Munich, the ingredients are delivered each day by the store-keeper, to the chief cook; and a person of confidence, not belonging to the kitchen, attends at the proper hour to see that they are actually used. Some one of the inspectors, or other chief officer of the establishment, also attends at the hour of dinner, to see that the victuals furnished to the poor are good; well dressed; and properly served up.

As the dining-hall is not large enough to accommodate all the poor at once, they dine in companies of as many as can be seated together, (about 150); those who work in the house being served first, and then those who come from the town.

Though most of those who work in their own lodgings send for their dinners, yet there are many others, and particularly such as from great age or other bodily infirmities are not able to work, who come from the town every day to the public hall to dine; and as these are frequently obliged to wait some time at the door, before they can be admitted into the dining-hall;—that is to say, till all the poor who work in the house have finished their dinners;—for their more comfortable accommodation, a large room, provided with a stove for heating it in winter, has been constructed, adjoining to the building of the institution, but not within the court, where these poor people assemble, and are sheltered from the inclemency of the weather while they wait for admittance into the dining-hall.

To preserve order and decorum at these public dinners, and to prevent crowding and jostling at the door of the dining-hall, the steward, or some other officer of the house of some authority, is always present in the hall during dinner; and two privates of the police guards, who know most of the poor personally, take post at the door of the hall, one on each side of it; and between them the poor are obliged to pass singly into the hall.

As soon as a company have taken places at the table, (the soup being always served out and placed upon the tables before they are admitted,) upon a signal given by the officer who presides at the dinner, they all repeat a short prayer. Perhaps I ought to ask pardon for mentioning so old-fashioned a custom; but I own I am old-fashioned enough myself to like such things.

As an account in detail will be given in another place, of the ex-pence of feeding these poor people, I shall only observe here, that this expense was considerably lessened by the voluntary donations of bread, and offal meat, which were made by the bakers and butchers of the town and suburbs. The beggars, not satisfied with the money which they extorted from all ranks of people by their unceasing importunity, had contrived to lay certain classes of the inhabitants under regular periodical contributions of certain commodities; and especially eatables; which they collected in kind. Of this nature were the contributions which were levied by them upon the bakers, butchers, keepers of eating-houses, ale-house keepers, brewers, etc. all of whom were obliged, at stated periods;—once a-week at least;—or oftener;— to deliver to such of the beggars as presented themselves at the hour appointed, very considerable quantities of bread, meat, soup, and other eatables; and to such a length were these shameful impositions carried, that a considerable traffic was actually carried on with the articles so collected, between the beggars, and a number of petty shop-keepers, or hucksters, who purchased them of the beggars, and made a business of selling them by retail to the indigent and industrious inhabitants. And though these abuses were well known to the public, yet this custom had so long existed, and so formidable were the beggars became to the inhabitants, that it was no means safe, or advisable, to refuse their demands.

Upon the town being cleared of beggars, these impositions ceased of course; and the worthy citizens, who were relieved from this burthen, felt so sensibly the service that was rendered them, that, to show their gratitude, and their desire to assist in supporting so useful an establishment, they voluntarily offered, in addition to their monthly subscriptions in money, to contribute every day a certain quantity of bread, meat, soup, etc. towards feeding the poor in the Military Work-house. And these articles were collected every day by the servants of the establishment; who went round the town with small carts, neatly fitted up, and elegantly painted, and drawn by single small horses, neatly harnessed.

As in these, as well as in all other collections of public charity, it was necessary to arrange matters so that the public might safely place the most perfect confidence in those who were charged with these details; the collections were made in a manner in which it was EVIDENTLY IMPOSSIBLE for those employed in making them to defraud the poor of any part of that which their charitable and more opulent fellow-citizens designed for their relief.—And to this circumstance principally it may, I believe, be attributed, that these donations have for such a length of time (more than five years,) continued to be so considerable.

In the collection of the soup, and the offal meat at the butchers' shops, as those articles were not very valuable and not easily concealed or disposed of, no particular precautions were necessary, other than sending round PUBLICLY and at a CERTAIN HOUR the carts destined for those purposes. Upon that for collecting the soup, which was upon four wheels, was a large cask neatly painted with an inscription on each side in large letters, "for the "Poor." That for the meat held a large tub with a cover, painted with the same colours, and marked on both sides with the same inscription.

Beside this tub, other smaller tubs, painted in like manner, and bearing the same inscription, "for the Poor," were provided and hung up in conspicuous situations in all the butchers' shops in the town. In doing this, two objects were had in view, first the convenience of the butchers; that in cutting up their meat they might have a convenient place to lay by that which they should destine for the poor till it should be called for; and secondly, to give an opportunity to those who bought meat in their shops to throw in any odd scraps, or bones, they might receive, and which they might not think worth the trouble of carrying home.

These odd pieces are more frequently to be met with in the lots which are sold in the butchers' shops in Munich than in almost any other town; for the price of meat is fixed by authority, the butchers have a right to sell the whole carcase, the bad pieces with the good, so that with each good lot there is what in this country is called the zugewicht, that is to say, an indifferent scrap of offal meat, or piece of bone, to make up the weight;— and these refuse pieces were very often thrown into the poor's tub; and after being properly cleaned and boiled, served to make their soup much more savoury and nourishing.

In the collection of the daily donations of bread, as that article is more valuable, and more easily concealed and disposed of, more precautions were used to prevent frauds on the parts of the servants who were sent round to make the collection.

The cart which was employed for this purpose was furnished with a large wooden chest, firmly nailed down upon it, and provided with a good lock and key; and this chest, which was neatly painted, and embellished with a inscription, was so contrived, by means of an opening in the top of a large vertical wooden tube fixed in its lid, and made in the form of a mouse-trap, that when it was locked, (as it always was when it was sent round for the donations of bread,) a loaf of bread, or any thing of that size, could be put into it; but nothing could be taken out of it by the same opening. Upon the return of the cart, the bread-chest was opened by the steward, who keeps the key of it; and its contents, after being entered in a register kept for that purpose, were delivered over to the care of the store-keeper.

The bread collected was commonly such as not having been sold in time, had become too old, hard, and stale for the market; but which, being cut fine, a handful of it put into a basin of good pease-soup, was a great addition to it.

The amount of these charitable donations in kind, may be seen in the transactions of the original returns, which are annexed in the Appendix, No. III.

The collections of soup were not long continued, it being found to be in general of much too inferior a quality to be mixed with the soup made in the kitchen of the poor-house; but the collections of bread, and of meat, continue to this time, and are still very productive.

But the greatest resource in feeding the poor, is one which I am but just beginning to avail myself of,—the use of potatoes[10]. Of this subject, however, I shall treat more largely hereafter.

The above-mentioned precautions used in making collections in kind, may perhaps appear trifling, and superfluous; they were nevertheless very necessary. It was also found necessary to change all the poor's-boxes in the churches, to prevent their being robbed; for though in those which were first put up, the openings were not only small, but ended in a curved tube, so that it appeared almost impossible to get any of the money out of the box by the same opening by which it was put into it; yet means were found, by introducing into the opening thin pieces of elastic wood, covered with bird-lime, to rob the boxes. This was prevented in the new boxes, by causing the money to descend through a sort of bag, with a hold in the bottom of it, or rather a flexible tube, made of chain-work, with iron wire, suspended in the middle of the box.

CHAPTER. VI.

Apology for the want of method in treating the subject under
 consideration.
Of the various means used for encouraging industry among the
poor.
Of the internal arrangement and government of the house of indus-
try.
Why called the military work-house.
Of the manner in which the business is carried on there.
Of the various means used for preventing frauds in carrying on the
 business in the different manufactures.
Of the flourishing state of those manufactures.

Though all the different parts of a well arranged establishment go
on together, and harmonize, like the parts of a piece of music in full
score, yet, in describing such an establishment, it is impossible to
write like the musician, in score, and to make all the parts of the
narrative advance together. Various movements, which exist together, and which have the most intimate connection and dependence upon each other, must nevertheless be described separately;
and the greatest care and attention, and frequently no small share of
address, are necessary in the management of such descriptions, to
render the details intelligible; and to give the whole its full effect of
order;—dependence;— connection;—and harmony. And in no case
can these difficulties be greater, than in descriptions like those in
which I am now engaged; where the number of the objects, and of
the details, is so great, that it is difficult to determine which should
be attended to first; and how far it may safely be pursued, without
danger of the others being too far removed from their proper places;—or excluded;— or forgotten.

The various measures adopted, and precautions taken, in arresting the beggars,—in collecting and distributing alms,—in establishing order and police among them,—in feeding and clothing the
poor,— and in establishing various manufactures for giving them
employment, are all subjects which deserve, and require, the most
particular explanation; yet those are not only operations which were

begun at the same time; and carried on together; but they are so dependent upon each other, that it is almost impossible to have a complete idea of the one, without being acquainted with the others; or of treating of the one, without mentioning the others at the same time. — This, therefore, must be my excuse, if I am taxed with want of method, or of perspicuity in the descriptions; and this being premised, I shall proceed to give an account of the various objects and operations which yet remain to be described.

I have already observed how necessary it was to encourage, by every possible means, a spirit of industry and emulation among those, who, from leading a life of indolence and debauchery, were to be made useful members of society; and I have mentioned some of the measures which were adopted for that purpose. It remains for me to pursue this interesting subject, and to treat it, in all its details, with that care and attention which its importance so justly demands.

Though a very generous price was paid for labour, in the different manufactures in which the poor were employed, yet, that alone was not enough to interest them sufficiently in the occupations in which they were engaged. To excite their activity, and inspire them with a true spirit of persevering industry, it was necessary to fire them with emulation; — to awaken in them a dormant passion, whose influence they had never felt; — the love of honest fame; — and ardent desire to excel; — the love of glory; — or by what other more humble or pompous name this passion, the most noble, and most beneficent that warms the human heart, can be distinguished.

To excite emulation; — praise; — distinctions; — rewards are necessary; and these were all employed. Those who distinguished themselves by their application, — by their industry, — by their address, — were publicly praised and encouraged; — brought forward, and placed in the most conspicuous situations; — pointed out to strangers who visited the establishment; and particularly named and proposed as models for others to copy. A particular dress, a sort of uniform for the establishment, which, though very economical, as may be seen by the details which will be given of it in another place, was nevertheless elegant, was provided; and this dress, as it was given out gratis, and only bestowed upon those who particularly distinguished themselves, was soon looked upon as an honourable mark of approved merit; and served very powerfully to excite emulation among the competitors, I doubt whether vanity, in any instance, ever surveyed itself with more self-gratification, than did some of these poor people when they first put on their new dress.

How necessary is it to be acquainted with the secret springs of action in the human heart, to direct even the lowest and most unfeeling class of mankind!—The machine is intrinsically the same in all situations;—the great secret is, FIRST TO PUT IT IN TUNE, before an attempt is made to play upon it. The jarring sounds of former vibrations must first be stilled, otherwise no harmony can be produced; but when the instrument is in order, the notes CANNOT FAIL to answer to the touch of a skilful master.

Though every thing was done that could be devised to impress the minds of all those, old and young, who frequented this establishment, with such sentiments as were necessary in order to their becoming good and useful members of society; (and in these attempts I was certainly successful, much beyond my most sanguine expectations;) yet my hopes were chiefly placed on the rising generation.

The children, therefore, of the poor, were objects of my peculiar care and attention. To induce their parents to send them to the establishment, even before they were old enough to do any kind of work, when they attended at the regular hours, they not only received their dinner gratis, but each of them was paid THREE CREUTZERS a day for doing nothing, but merely being present where others worked.

I have already mentioned that these children, who were too young to work, were placed upon seats built round the halls where other children worked. This was done in order to inspire them with a desire to do that, which other children, apparently more favoured, —more caressed,—and more praised than themselves, were permitted to do; and of which they were obliged to be idle spectators; and this had the desired effect.

As nothing is so tedious to a child as being obliged to sit still in the same place for a considerable time, and as the work which the other more favoured children were engaged in, was light and easy, and appeared rather amusing than otherwise, being the spinning of hemp and flax, with small light wheels, turned with the foot, these children, who were obliged to be spectators of this busy and entertaining scene, became so uneasy in their situations, and so jealous of those who were permitted to be more active, that they frequently solicited with the greatest importunity to be permitted to work, and often cried most heartily if this favour was not instantly granted them.

How sweet these tears were to me, can easily be imagined!

The joy they showed upon being permitted to descend from their benches, and mix with the working children below, was equal to the solicitude with which they had demanded that favour.

They were at first merely furnished with a wheel, which they turned for several days with the foot, without being permitted to attempt any thing further. As soon as they were become dexterous in the simple operation, and habit had made it so easy and familiar to them that the foot could continue its motion mechanically, without the assistance of the head;—till they could go on with their work, even though their attention was employed upon something else;—till they could answer questions, and converse freely with those about them upon indifferent subjects, without interrupting or embarrassing the regular motion of the wheel, then,—and not till then,—they were furnished with hemp or flax, and were taught to spin.

When they had arrived at a certain degree of dexterity in spinning hemp and flax, they were put to spinning of wool; and this was always represented to them, and considered by them, as an honorable promotion. Upon this occasion they commonly received some public reward, a new shirt,—a pair of shoes,— or perhaps the uniform of the establishment, as an encouragement to them to persevere in their industrious habits.

As constant application to any occupation for too great a length of time is apt to produce disgust, and in children might even be detrimental to health, beside the hour of dinner, an hour of relaxation from work, (from eight o'clock till nine,) in the forenoon, and another hour, (from three o'clock till four,) in the afternoon, were allowed them, and these two hours were spent in a school; which, for want of room elsewhere in the house, was kept in the dining-hall, where they were taught reading, writing, and arithmetic, by a school-master engaged and paid for that purpose[11]. Into this school other persons who worked in the house, of a more advanced age, were admitted, if they requested it; but few grown persons seemed desirous of availing themselves of this permission. As to the children, they had no choice in the matter; those who belonged to the establishment were obliged to attend the school regularly every day, morning and evening. The school books, paper, pens, and ink, were furnished at the expence of the establishment.

To distinguish those among the grown persons that worked in the house, who showed the greatest dexterity and industry in the different manufactures in which they were employed, the best workman were separated from the others, and formed distinct classes, and were even assigned separate rooms and apartments. This separation was productive of many advantages; for, beside the spirit of emulation which it excited, and kept alive, in every part of the establishment, if afforded an opportunity of carrying on the different manufactures in a very advantageous manner. The most dexterous among the wool-spinners, for instance, were naturally employed upon the finest wool, such as was used in the fabrication of the finest and most valuable goods; and it was very necessary that these spinners should be separated from the others, who worked upon coarser materials; otherwise, in the manipulations of the wool, as particles of it are unavoidably dispersed about in all directions when it is spun, the coarser particles thus mixing with the fine would greatly injure the manufacture. It was likewise necessary, for a similar reason, to separate the spinners who were employed in spinning wool of different colours. But as these, and many other like precautions are well known to all manufacturers, it is not necessary that I should insist upon them any farther in this place; nor indeed is it necessary that I should enter into all the details of any of the manufactures carried on in the establishment I am describing. It will be quite sufficient, if I merely enumerate them, and others, who were employed in carrying them on.

In treating this subject it will however be necessary to go back a little, and give a more particular account of the internal governments of this establishment; and first of all I must observe, that the government of the Military Work-house, as it is called, is quite distinct from the government of the institution for the poor; the Workhouse being merely a manufactory, like any other manufactory, supported upon its own private capital; which capital has no connection whatever with any fund destined for the poor. It is under the sole direction of its own particular governors and overseers, and is carried on at the sole risk of the owner. The institution for the poor, on the other hand, is merely an institution of charity, joined to a general direction of the police, as far as it relates to paupers. The committee, or deputation, as it is called, which is at the head of this institution, has the sole direction of all funds destined for the relief of the poor in Munich, and the distribution of alms. This deputation has likewise the direction of the kitchen, and bake-house, which are established in the Military Work-house; and of the details relative to the feeding of the poor; for it is from the funds destined for the relief of the poor that these expences are defrayed: the deputation is also in connection with the Military Work-house relative to the clothing of the poor, and the distribution of rewards to those of them who particularly distinguished themselves by their good behaviour and their industry, but this is merely a mercantile correspondence. The deputation has no right to interfere in any way whatever in the internal management of this establishment, considered as a manufactory. In this respect it is to all intents and purposes a perfectly distinct and independent establishment. But notwithstanding this, the two establishments are so dependent on each other in many respects, that neither of them could well subsist alone.

The Military Work-house being principally designed as a manufactory for clothing the army, its capital, which at first consisted in about 150,000 florins, but which has since increased to above 250,000 florins, was advanced by the military chest, and hence it is, that it was called the Military Work-house, and put under the direction of the council of war.

For the internal management of the establishment, a special commission was named, consisting of, one counsellor of war, of the department of military economy, or of the clothing of the army,—one captain, which last is inspector of the house, and has apartments in it, where he lodges; —and the store-keeper of the magazine of military clothing.

These commissioners, who have the magazine of military clothing at the same time under their direction, have, under my immediate superintendence, the sole government and direction of this establishment;—of all the inferior officers;—servants;— manufacturers;—and workmen, belonging to it; and of all mercantile operations;—contracts;— purchases;—sales;, etc. And it is with these commissioners that the regiments correspond, in order to be furnished with clothing, and other necessaries; and into their hands they pay the amount of the different articles received.

The cash belonging to this establishment is placed in a chest furnished with three separate locks, of one of which each of the commissioners are jointly, and severally, answerable for the contents of the chest.

These commissioners hold their sessions regularly twice a week, or oftener if circumstances require it, in a room in the Military Work-house destined for that purpose, where the correspondence, and all accounts and documents belonging to the establishment, and other records, are kept; and where the secretary of the commission constantly attends.

When very large contracts are made for the purchase of raw materials, particularly when they are made with foreigners, the conditions are first submitted by the commissioners to the council of war for their approbation; but in all concerns of less moment, and particularly in all the current business of the establishment;—in the ordinary purchases,—sales,—and other mercantile transactions; the commissioners act by their own immediate authority: but all the transactions of the commissioners BEING ENTERED REGULARLY IN THEIR JOURNALS, and the most particular account of all sales, and purchases, and other receipts and expenditures being kept; and inventories being taken every year, of all raw materials;—manufactures upon hand;—and other effects, belonging to the establishment; and an annual account of profit and loss, regularly made out; all peculation, and other abuses, are most effectually prevented.

The steward, or store-keeper of raw materials, as he is called, has the care of all raw materials, and of all finished manufactures destined for private sale. The former are kept in magazines, or store-rooms, of which he alone has the keys,— the latter are kept in rooms set apart as a store,—or shop,— where they are exposed for public inspection, and sale. To prevent abuses in the sales of these manufactures, their prices, which are determined upon a calculation of what they cost, and a certain per cent. added for the profits of the house, are marked upon the goods, and are never altered; and a regular account is kept of all, even of the most inconsiderable articles sold, in which not only the commodity, with its quality, quantity, an price, is specified; but the name of the purchaser, and the day of the month when the purchase was made, are mentioned.

All articles of clothing destined for the army which are made up in the house; as well as all goods in the piece, destined for military clothing, are lodged in the Military Magazine; which is situated at some distance from the Military Work-house; and is under the care and inspection of the Military store-keeper.

From this Military Magazine, which may be considered as an appendix to the Military Work-house, and is in fact under the same direction, the regiments are supplied with every article of their clothing. But in order that the army accounts may be more simple, and more easily checked, and that the total annual expence of each regiment may be more readily ascertained, the regiments pay, at certain fixed prices, for all the articles they receive from the Military Magazine, and charge such expenditures in the annual account which they send in to the War Office.

The order observed with regard to the delivery of the raw materials by the store-keeper or steward of the Military Work-house to those employed in manufacturing them, is as follows:

In the manufactures of wool, for instance, he delivers to the master-clothier a certain quantity, commonly 100 pounds, of wool, of a certain quality and description; taken from a certain division, or bin, in the Magazine; bearing a certain number; in order to its being sorted. And as a register is kept of the wool that is put into these bins from time to time, and as the lots of wool are always kept separate, it is perfectly easy at any time to determine when,—and where,—and from whom, the wool delivered to the sorted was purchased; and what was paid for it; and consequently, to trace the wool from the stock where it was grown, to the cloth into which it was formed; and even to the person who wore it. And similar arrangements are adopted with regard to all other raw materials used in the various manufactures.

The advantages arising from this arrangement are too obvious to require being particularly mentioned. It not only prevents numberless abuses on the part of those employed in the various manufactures, but affords a ready method of detecting any frauds on the part of those from whom the raw materials are purchased.

The wool received by the master-clothier is by him delivered to the wool-sorters to be sorted. To prevent frauds on the part of the wool-sorters, not only all the wool-sorters work in the same room, under the immediate inspection of the master wool-sorter, but a certain quantity of each lot of wool being sorted in the presence of some one of the public officers belonging to the house, it is seen by the experiment how much per cent. is lost by separation of dirt and filth in sorting; and the quantity of sorted wool of the different qualities, which the sorter is obliged to deliver for each HUNDRED POUNDS weight of wool received from the magazine, is from hence determined.

The great secret of the woollen manufactory is in the sorting of the wool, and if this is not particularly attended to; that is to say, if the different kinds of wool of various qualities which each fleece naturally contains, are not carefully separated; and if each kind of wool is not employed for that purpose, and FOR THAT ALONE, for which it is best calculated, no woollen manufactory can possibly subsist with advantages.

Each fleece is commonly separated into five or six different parcels of wool, of different qualities, by the sorters in the Military Work-house; and of these parcels, some are employed for warp;—others for wool;—others for combing;—and that which is very coarse and indifferent, for coarse mittens for the peasants;—for the lists of broad cloths, etc.

The wool, when sorted, is delivered back by the master-clothier to the steward, who now places it in the sorted-wool magazines, where it is kept in separate bins, according to its different qualities and destinations, till it is delivered out to be manufactured. As these bins are all numbered, and as the quality and destination of the wool which is lodged in each bin is always the same, it is sufficient in describing the wool afterwards as it passes through the hands of the different manufacturers, merely to mention ITS NUMBER; that is to say, the number of the bin the sorted-wool magazine from whence it was taken.

As a more particular account of these various manipulations, and the means used to prevent frauds, may not only be interesting to all who are curious in these matters, but may also be of real use to such as may engage in similar undertakings, I shall take the liberty to enlarge a little upon this subject.

From the magazine of sorted wool, the master-clothier receives this sorted wool again, in order to its being wolfed, — greased, — carded; — and spun, under his inspection, and then delivered into the store-room of woollen yarn. As woollen yarn he receives it again, and delivers it to the cloth-weaver. — The cloth-weaver returns it in cloth to the steward. — The steward delivers it to the fuller; — the fuller to the cloth-shearer; — the cloth-shearer to the cloth-presser; — and the cloth-presser to the steward; — and by this last it is delivered into the Military Magazine, if destined for the army; if not, it is placed in the shop for sale. The master-clothier is answerable for all the sorted wool he receives, till he delivers it to the clerk of the wool-spinners; and all his accounts are settled with the steward once a week. — The clerk of the spinners is answerable for the carded and combed wool he receives from the master-clothier, till it is delivered in yarn in the store-room; and his accounts are likewise settled with the master-clothier, and with the clerk of the store-room, (who is called the clerk of the control,) once a week. The spinners wages are paid by the clerk of the control, upon the spin-ticket, signed by the clerk of the spinners; in which ticket, the quantity, and quality of the yarn spun being specified, together with the name of the spinner, the weekly delivery of yarn by the clerk of the spinners into the store-room, must answer to the spin-tickets received and paid by the clerk of the control. More effectually to prevent frauds, each delivery of yarn to the clerk of the spinners is bound up in a separate bundle, to which is attached an abstract of the spin-ticket, in which abstract is specified, the name of the spinner;—the date of the delivery;—the number of the spin-ticket;—and the quantity and quality of the yarn. This arrangement not only facilitates the settlement of the weekly account between the clerk of the spinners and the clerk of the control, when the former makes his weekly delivery of yarn into the store-room, but renders it easy also to detect any frauds committed by the spinners.

The wages of the spinners are regulated by the fineness of the yarn; that is, by the number of skains, or rather knots, which they spin from the pound of wool. Each knot is composed of 100 threads, and each thread, or turn of the reel, is two Bavarian yards in length; and to prevent frauds in reeling, clock-reels, proved and sealed, are furnished by the establishment to all the spinners. It is possible, however, notwithstanding this precaution, for the spinners to commit frauds, by binding up knots containing a smaller number of threads than 100.—It is true they have little temptation to do so, for as their wages are in fact paid by the WEIGHT of the yarn delivered, and the number of knots serving merely to determine the price BY THE POUND which they have a right to receive, and advantages they can derive from frauds committed in reeling are very trifling indeed. But trifling as they are, such frauds would no doubt sometimes be committed, were it not known that it is absolutely IMPOSSIBLE for them to escape detection.

Not only the clerk of the spinners examines the yarn when he receives it, and counts the threads in any of the knots which appear to be too small, but the name of the spinner, with a note of the quantity of knots, accompanies the yarn into the store-room, as was before observed, and from thence to the spooler, by whom it is wound off; any frauds committed in reeling cannot fail to be brought home to the spinner.

The bundles of carded wool delivered to the spinners, though they are called pounds, are not exact pounds. They contain each as much more than a pound, as is necessary, allowing for wastage in spinning, in order that the yarn when spun may weigh a pound. If the yarn is found to be wanting in weight, a proportional deduction is made from the wages of the spinner; which deduction, to prevent frauds, amounts to a trifle more than the value of the yarn which is wanting.

Frauds in weaving are prevented by delivering the yarn to the weavers by weight, and receiving the cloth by weight from the loom. In the other operations of the manufactures, such as fulling, shearing, pressing, etc. no frauds are to be apprehended.

Similar precautions are taken to prevent frauds in the linen;—cotton;—and other manufactures carried on in the house; and so effectual are the means adopted, that during more than five years since the establishment was instituted, no one fraud of the least consequence has been discovered; the evident impossibility of escaping detection in those practices, having prevented the attempt.

Through the above-mentioned details may be sufficient to give some idea of the general order which reigns in every part of this extensive establishment; yet, as success in an undertaking of this kind depends essentially on carrying on the business in all its various branches in the most methodical manner, and rendering one operation a check upon the other, as well as in making the persons employed absolutely responsible for all frauds and neglects committed in their various departments, I shall either add in the Appendix, or publish separately, a full account of the internal details of the various trades and manufactures carried on in the Military Work-house, and copies of all the different tickets,—returns,—tables,—accounts, etc. made use of in carrying on the business of this establishment.

Though these accounts will render this work more voluminous than I could have wished, yet, as such details can hardly fail to be very useful to those, who, either upon a larger, or smaller scale, may engage in similar undertakings, I have determined to publish them.

To show that the regulations observed in carrying on the various trades and manufactures in the Military Work-house are good, it will, I flatter myself, be quite sufficient to refer to the flourishing state of the establishment;—to its growing reputation;—to its extensive connections, which reach even to foreign countries;—to the punctuality with which all its engagements are fulfilled;— to its unimpeached credit;—and to its growing wealth.

Notwithstanding all the disadvantages under which it laboured in its infant state, the net profits arising from it during the six years it has existed, amount to above 100,000 florins; after the expences of every kind,—salaries,—wages,—repairs, etc. have been deducted; in consequence of the augmentation of the amount of the orders received and executed the last year, did not fall much short of HALF A MILLION of florins.

It may be proper to observe, that, not the whole army of the Elector, but only the fifteen Bavarian regiments, are furnished with clothing from the Military Work-house at Munich. The troops of the Palatinate, and those of the Duchies of Juliers and Bergen, receive their clothing from a similar establishment at Manheim.

The Military Work-house at Manheim was indeed erected several months before that at Munich; but as it is not immediately connected with any institution for the poor,—as the poor are not fed in it,—and as it was my first attempt, or coup d'essai,— it is, in many respects, inferior in its internal arrangements to that at Munich. I have therefore chosen this last for the subject of my descriptions; and would propose it as a model for imitation, in preference to the other.

As both these establishments owe their existence to myself, and as they both remain under my immediate superintendence, it may very naturally be asked, why that at Manheim has not been put upon the same footing with that at Munich?—My answer to this question would be, that a variety of circumstances, too foreign to my present subject to be explained here, prevented the establishment of the Military Work-house at Manheim being carried to that perfection which I could have wished[12].

But it is time that I should return to the poor of Munich; for whose comfort and happiness I laboured with so much pleasure, and whose history will ever remain by far the most interesting part of this publication.

CHAPTER. VII.

A further account of the poor who were brought together in the
 house of industry:—and of the interesting change which was
 produced in their manners and dispositions.
Various proofs that the means used for making them industrious,
 comfortable, and happy, were successful.

The awkwardness of these poor creatures, when they were first
taken from the streets as beggars, and put to work, may easily conceived; but the facility with which they acquired address in the
various manufactures in which they were employed, was very remarkable, and much exceeded my expectation. But what was quite
surprising, and at the same time interesting in the highest degree,
was the apparent and rapid change which was produced in their
manners,—in their general behaviour,—and even in the very air of
their countenances, upon being a little accustomed to their new
situations. The kind usage they met with, and the comforts they
enjoyed, seemed to have softened their hearts, and awakened in
them sentiments as new and surprising to themselves, as they were
interesting to those about them.

The melancholy gloom of misery, and air of uneasiness and embarrassment, disappeared by little and little from their countenances, and were succeeded by a timid dawn of cheerfulness, rendered most exquisitely interesting by a certain mixture of silent
gratitude, which no language can describe.

In the infancy of this establishment, when these poor creatures
were first brought together, I used very frequently to visit them,—to
speak kindly to them,—and to encourage them;—and I seldom
passed through the halls where they were at work, without being a
witness to the most moving scenes.

Objects, formerly the most miserable and wretched, whom I had
seen for years as beggars in the streets;-young women,—perhaps
the unhappy victims of seduction, who, having lost their reputation,
and being turned adrift in the world, without a friend and without a
home, were reduced to the necessity of begging, to sustain a miser-

able existence, now recognized me as their benefactor; and, with tears dropping fast from their cheeks, continued their work in the most expressive silence.

If they were asked, what the matter was with them? their answer was, ("nichts") "nothing;" accompanied by a look of affectionate regard and gratitude, so exquisitely touching as frequently to draw tears from the most insensible of the bystanders.

It was not possible to be mistaken with respect to the real state of the minds of these poor people; every thing about them showed that they were deeply affected with the kindness shown them;— and that their hearts were really softened, appeared, not only from their unaffected expressions of gratitude, but also from the effusions of their affectionate regard for those who were dear to them. In short, never did I witness such affecting scenes as passed between some of these poor people and their children.

It was mentioned above that the children were separated from the grown persons. This was the case at first; but as soon as order was thoroughly established in every part of the house, and the poor people had acquired a certain degree of address in their work, and evidently took pleasure in it, as many of those who had children expressed an earnest desire to have them near them, permission was granted for that purpose; and the spinning halls, by degrees, were filled with the most interesting little groups of industrious families, who vied with each other in diligence and address; and who displayed a scene, at once the most busy, and the most cheerful, that can be imagined.

An industrious family is ever a pleasing object; but there was something peculiarly interesting and affecting in the groups of these poor people. Whether it was, that those who saw them compared their present situation with the state of misery and wretchedness from which they had been taken; —or whether it was the joy and exultation which were expressed in the countenances of the poor parents in contemplating their children all busily employed about them;—or the air of self-satisfaction which these little urchins put on, at the consciousness of their own dexterity, while they pursued their work with redoubled diligence upon being observed, that rendered the scene so singularly interesting,— I know not; but certain it is, that few strangers who visited the establishment, came out of these halls without being much affected.

Many humane and well-disposed persons are often withheld from giving alms, on account of the bad character of beggars in general; but this circumstance, though it ought undoubtedly to be taken into consideration in determining the mode of administering our charitable assistance, should certainly not prevent our interesting ourselves in the fate of these unhappy beings. On the contrary, it ought to be an additional incitement to us to relieve them;—for nothing is more certain, than that their crimes are very often the EFFECTS, not the CAUSES of their misery; and when this is the case, by removing the cause, the effects will cease.

Nothing is more extraordinary and unaccountable, than the inconsistency of mankind in every thing; even in the practice of that divine virtue benevolence; and most of our mistakes arise more from indolence and from inattention, than from any thing else. The busy part of mankind are too intent upon their own private pursuits; and those who have leisure, are too averse from giving themselves trouble, to investigate a subject but too generally considered as tiresome and uninteresting. But if it be true, that we are really happy only in proportion as we ought to be so;— that is, in proportion as we are instrumental in promoting the happiness of others; no study surely can be so interesting, as that which teaches us how most effectually to contribute to the well-being of our fellow-creatures.

If LOVE be blind, SELF-LOVE is certainly very short-sighted; and without the assistance of reason and reflection, is but a bad guide in the pursuit of happiness.

Those who take pleasure in depreciating all the social virtues have represented pity as a mere selfish passion; and there are some circumstances which appear to justify this opinion. It is certain that the misfortunes of others affect us, not in proportion to their greatness, but in proportion to their nearness to ourselves; or to the chances that they may reach us in our turns. A rich man is infinitely more affected at the misfortune of his neighbour, who, by the failure of a banker with whom he had trusted the greater part of his fortune;—by an unlucky run at play,—or by other losses, is reduced to a state of affluence, to the necessity of laying down his carriage;—leaving the town;—and retiring into the country upon a few hundreds a-year;—than by the total ruin of the industrious tradesman over the way, who is dragged to prison, and his numerous family of young and helpless children left to starve.

But however selfish pity may be, BENEVOLENCE certainly springs from a more noble origin. It is a good-natured,—generous sentiment, which does not require being put to the torture in order to be stimulated to action. And it is this sentiment, not pity, or compassion, which I would wish to excite.

Pity is always attended with pain; and if our sufferings at being witnesses of the distresses of others, sometimes force us to relieve them, we can neither have much merit, nor any lasting satisfaction, from such involuntary acts of charity; but the enjoyments which result from acts of genuine benevolence are as lasting as they are exquisitely delightful; and the more they contribute to that inward peace of mind and self-approbation, which alone constitute real happiness. This is the "soul's calm sun-shine, and the heart-felt joy," which is virtue's prize.

To induce mankind to engage in any enterprise, it is necessary, first, to show that success will be attended with real advantage; and secondly, that is may be obtained without much difficulty. The rewards attendant upon acts of benevolence have so often been described and celebrated, in every country and in every language, that it would be presumption in me to suppose I could add any thing new upon a subject already discussed by the greatest masters of rhetoric, and embellished with all the irresistible charms of eloquence; but as EXAMPLE OF SUCCESS are sometimes more efficacious in stimulating mankind to action, than the most splendid reasonings and admonitions, it is upon my SUCCESS in the enterprise of which I have undertaken to give an account, that my hopes of engaging others to follow such an example are chiefly founded; and hence it is, that I so often return to that part of my subject, and insist with so much perseverance upon the pleasure which this success afforded me. I am aware that I expose myself to being suspected of ostentation, particularly by those who are not able to enter fully into my situation and feelings; but neither this, nor any other consideration, shall prevent me from treating the subject in such a manner as may appear best adapted to render my labours of public utility.

Why should I not mention even the marks of affectionate regard and respect which I received from the poor people for those happiness I interested myself, and the testimonies of the public esteem with which I was honored?—Will it be reckoned vanity, if I mention the concern which the Poor of Munich expressed in so affecting a manner when I was dangerously ill?—that they went publicly in a body in procession to the cathedral church, where they had divine service performed, and put up public prayers for my recovery?—that four years afterwards, on hearing that I was again dangerously ill at Naples. they, of their own accord, set apart an hour each evening, after they had finished their work in the Military Work-house, to pray for me?

Will it be thought improper to mention the affecting reception I met with from them, at my first visit to the Military Work-house upon my return to Munich last summer, after an absence of fifteen months; a scene which drew tears from all who were present?—and must I refute myself the satisfaction of describing the fete I gave them in return, in the English Garden, at which 1800 poor people of all ages, and above 30,000 of the inhabitants of Munich, assisted? and all this pleasure I must forego, merely that I may not be thought vain and ostentatious?—Be it so then;— but I would just beg leave to call the reader's attention to my feelings upon the occasion; and then let him ask himself, if any earthly reward can possibly be supposed greater;—any enjoyments more complete, than those I received. Let him figure to himself, if he can, my situation, sick in bed, worn out by intense application, and dying, as every body thought, a martyr in the cause to which I had devoted myself;—let him imagine, I say, my feelings, upon hearing the confused noise of the prayers of a multitude of people, who were passing by in the streets, upon being told, that it was the Poor of Munich, many hundreds in number, who were going in procession to the church to put up public prayers for me:—public prayers for me!—for a private person!— a stranger!—a protestant!—I believe it is the first instance of the kind that ever happened;—and I dare venture to affirm that no proof could well be stronger than this, that the measures adopted for making these poor people happy, were really successful;—and let it be remembered, that this fact is what I am most anxious to make appear, IN THE CLEAREST AND MOST SATISFACTORY MANNER.

CHAPTER. VIII.

Of the means used for the relief of those poor persons who were
  not beggars.
Of the large sums of money distributed to the poor in alms.
Of the means used for rendering those who received alms industrious.
Of the general utility of the house of industry to the poor,
  and the distressed of all denominations.
Of public kitchens for feeding the poor, united with establishments
  for giving them employment; and of the great advantages which
  would be derived from forming them in every parish.
Of the manner in which the poor of Munich are lodged.

In giving an account of the Poor of Munich. I have hitherto confined myself chiefly to one class of them,—the beggars; but I shall
now proceed to mention briefly the measures which were adopted
to relieve others, who never were beggars, from those distresses and
difficulties in which poverty and the inability to provide the necessaries of life had involved them.

An establishment for the Poor should not only provide for the relief and support of those who are most forward and clamorous in
calling out for assistance;—humanity and justice require that peculiar attention should be paid to those who are bashful and silent.—
To those, who, in addition to all the distresses arising from poverty
and want, feel, that is still more insupportable to their unfortunate
and hopeless situation.

All those who stood in need of assistance were invited and encouraged to make known their wants to the committee placed at the
head of the institution; and in no case was the necessary assistance
refused.—That this relief was generously bestowed, will not be
doubted by those who are informed that the sums distributed in
alms, IN READY MONEY to the Poor of Munich in FIVE YEARS,
exclusive of the expences incurred in feeding and clothing them,
amounted to above TWO HUNDRED THOUSAND FLORINS[13].

But the sums of money distributed among the Poor in alms was not the only, and perhaps not the most important assistance that was given them.—THEY WERE TAUGHT AND ENCOURAGED TO BE INDUSTRIOUS; and they probably derived more essential advantages from the fruits of their industry, than from all the charitable donations they received.

All who were able to earn any thing by their labour, were furnished with work, and effectual measures taken to excite them to be industrious.—In fixing the amount of the sums in money, which they receive weekly upon stated days, care was always taken to find out how much the person applying for relief was in a condition to earn, and only just so much was granted, as, when added to these earnings, would be sufficient to provide the necessaries of life, or such of them as were not otherwise furnished by the institution. — But even this precaution would not alone have been sufficient to have obliged those who were disposed to be idle, to become industrious; for, with the assistance of the small allowances which were granted, they might have found means, by stealing, or other fraudulent practices, to have subsisted without working, and the sums allowed them would have only served as an encouragement to idleness.—This evil, which is always much to be apprehended in establishments for the Poor, and which is always most fatal in its consequences, is effectually prevented at Munich by the following simple arrangement:—A long and narrow slip of paper, upon which is printed, between parallel lines, in two or more columns, all the weeks in the year, or rather the month, and the day of the month, when each week begins, is, in the beginning of every year, given to each poor perform entitled to receive alms; and the name of the person,—with the number his name bears in the general list of the Poor;—the weekly sum granted to him,—and the sum he is able to earn weekly by labour, are entered in writing at the head of this list of the weeks.—This paper, which must always be produced by the poor person as often as he applies for his weekly allowance of alms, serves to show whether he has, or has not fulfilled the conditions upon which the allowance was granted him;— that is to say, whether he has been industrious, and has earned by his labour, and received, the sum he ought to earn weekly.— This fact is ascertained in the following manner: when the poor person frequents the house of industry regularly, or when he works at home, and delivers regularly at the end of every week, the produce of the labour he is expected to perform; when he has thus fulfilled the conditions imposed on him, the column, or rather parallel, in his paper, (which may be called his certificate of industry,) answering to the week in question, is marked with a stamp, kept for that purpose at the Military Work-house; or, if he should be prevented by illness, or any other accident, from fulfilling those conditions, in that case, instead of the stamp, the week must be marked by the signature of the commissary of the district to which the poor person belongs. — But, if the certificate be not marked, either by the stamp of the house of industry, or by the signature of the commissary of the district, the allowance for the week in question is not issued.

It is easy to be imagined how effectually this arrangement must operate as a check to idleness. — But, not satisfied with discouraging and punishing idleness, we have endeavoured, by all the means in our power, and more especially by rewards and honorable distinctions of every kind, to encourage extraordinary exertions of industry. Such of the Poor who earn more in the week than the sum imposed on them, are rewarded by extraordinary presents, in money, or in some useful and valuable article of clothing; or they are particularly remembered at the next public distribution of money, which is made twice a year to the Poor, to assist them in paying their house-rent: and so far is this from being made a pretext for diminishing their weekly allowance of alms, that it is rather considered as a reason for augmenting them.

There are great numbers of persons, of various descriptions, in all places, and particularly in great towns, who, though they find means just to support life, and have too much feeling ever to submit to the disgrace of becoming a burthen upon the public, are yet very unhappy, and consequently objects highly deserving of the commiseration and friendly aid of the humane and generous. — it is hardly possible to imagine a situation more truly deplorable than that of a person born to better prospects, reduced by unmerited misfortunes to poverty, and doomed to pass his whole life in one continued and hopeless struggle with want, shame, and despair.

Any relief which it is possible to afford to distress that appears under this respectable and most interesting form, ought surely never to be withheld.—But the greatest care and precaution are necessary in giving assistance to those who have been rendered irritable and suspicious by misfortunes, and who have too much honest pride not to feel themselves degraded by accepting an obligation they never can hope to repay.

The establishment of the house of industry at Munich has been a means of affording very essential relief to many distressed families, and single persons in indigent circumstances, who, otherwise, most probably never would have received any assistance. —Many persons of distinguished birth, and particularly widows and unmarried ladies with very small fortunes, frequently send privately to this house for raw materials,—flax or wool, — which they spin, and return in yarn,—linen for soldiers shirts, which they make up, etc. and receive in money, (commonly through the hands of a maid-servant, who is employed as a messenger upon these occasions,) the amount of the wages at the ordinary price paid by the manufactory, for the labour performed.

Many a common soldier in the Elector's service wears shirts made up privately by the delicate hands of persons who were never seen publicly to be employed in such coarse work;—and many a comfortable meal has been made in the town of Munich, in private, by persons accustomed to more sumptuous fare, upon the soup destined for the Poor, and furnished gratis from the public kitchen of the house of industry. Many others who stand in need of assistance, will, in time, I hope, get the better of their pride, and avail themselves of these advantages.

To render this establishment for the Poor at Munich perfect, something is still wanting.—The house of industry is too remote from the center of the town, and many of the Poor live at such a distance from it, that much time is lost in going and returning. —It is situated, it is true, nearly in the center of the district in which most of the Poor inhabit, but still there are many who do not derive all the advantages from it they otherwise would do were it adjacent to their dwellings. The only way to remedy this imperfection would be, to establish several smaller public kitchens in different parts of the town, with two or three rooms adjoining to each, where the Poor might work.—They might then either fetch the raw materials from the principal house of industry, or be furnished with them by the persons who superintend those subordinate kitchens; and who might serve at the same time as stewards and inspectors of the working rooms, under the direction and control of the officers who are placed at the head of the general establishment. This arrangement is in contemplation, and will be put in execution as soon as convenient houses can be procured and fitted up for the purpose.

In large cities, these public kitchens, and rooms adjoining to them for working, should be established in every parish; and, it is scarcely to be conceived how much this arrangement would contribute to the comfort and contentment of the Poor, and to the improvement of their morals. These working rooms might be fitted up with neatness; and even with elegance; and made perfectly warm, clean, and comfortable, at a very small expence; and, if nothing were done to disgust the Poor, either by treating them harshly, or using FORCE to oblige them to frequent these establishments, they would soon avail themselves of the advantages held out to them; and the tranquillity they would enjoy in these peaceful retreats, would, by degrees, calm the agitation of their minds,—remove their suspicions,—and render them happy,—grateful, and docile.

Though it might not be possible to provide any other lodgings for them than the miserable barracks they now occupy, yet, as they might spend the whole of the day, from morning till late at night, in these public rooms, and have no occasion to return to their homes till bed-time, they would not experience much inconvenience from the badness of the accommodation at their own dwellings.

Should any be attached with sickness, they might be sent to some hospital, or rooms be provided for them, as well as for the old and infirm, adjacent to the public working rooms. Certain hours might also be set apart for instructing the children, daily, in reading and writing, in the dining-hall, or in some other room convenient for that purpose.

The expence of forming such an establishment in every parish would not be great, in the first outset, and the advantages derived from it would very soon repay that expence, with interest. —The

Poor might be fed from a public kitchen for LESS THAN HALF what it would cost them to feed themselves;—they would turn their industry to better account, by working in a public establishment, and under proper direction, than by working at home;—a spirit of emulation would be excited among them, and they would pass their time more agreeably and cheerfully.— They would be entirely relieved from the heavy expense of fuel for cooking; and, in a great measure, from that for heating their dwellings; and, being seldom at home in the day-time, would want little more than a place to sleep in; so that the expence of lodging might be greatly diminished.—It is evident, that all these saving together would operate very powerfully to lessen the public expence for the maintenance of the Poor; and, were proper measures adopted, and pursued with care and perseverance, I am persuaded the expence would at last be reduced to little or nothing.

With regard to the lodgings for the Poor, I am clearly of opinion that it is in general best, particularly in great towns, that these should be left for themselves to provide. This they certainly would like better than being crowded together, and confined like prisoners in poor-houses and hospitals; and I really think the difference in the expence would be inconsiderable; and though they might be less comfortably accommodated, yet the inconvenience would be amply compensated BY THE CHARMS WHICH LIBERTY DISPENSES.

In Munich, almost all the Poor provide their own lodgings; and twice a year have certain allowances in money, to assist them in paying their rent.—Many among them who are single, have indeed, no lodgings they can call their own. They go to certain public-houses to sleep, where they are furnished with what is called a bed, in a garret, for one creutzer, (equal to about one-third of a penny,) a-night; and for two creutzers a-night they get a place in a tolerably good bed in a decent room in a public-house of more repute.

There are, however, among the Poor, many who are infirm, and not able to shift for themselves in the public-houses, and have not families, or near relations, to take care of them. For these, a particular arrangement has lately been made at Munich. Such of them as have friends or acquaintances in town with whom they can lodge, are permitted to do so; but if they cannot find out lodgings themselves, they have the option, either to be placed in some private family to be taken care of, or go to a home which has lately been purchased and fitted up as an hospital for lodging them[14].

This house is situated in a fine airy situation, on a small eminence upon the banks of the Isar, and overlooks the whole of the town;—the plain in which it is situated;—and the river.— It is neatly built, and has a spacious garden belonging to it. There are seventeen good rooms in the house; in which it is supposed about eighty persons may be lodged. These will all be fed from one kitchen; and such of them who are very infirm, will have others less infirm placed in the same room with them, to assist them, and wait upon them.—The cultivation of the garden will be their amusement, and the produce of it their property. —They will be furnished with work suitable to their strength; and for all the labour they perform, will be paid in money, which will be left at their own disposal.—They will be furnished with food, medicine, and clothing, gratis; and to those who are not able to earn any thing by labour, a small sum of money will be given weekly, to enable them to purchase tobacco, snuff, or any other article of humble luxury to which they may have been accustomed.

I could have wished that this asylum had been nearer to the house of industry. It is indeed not very far from it, perhaps not more than 400 yards; but still that is too far.—Had it been under the same roof, or adjoining to it, those who are lodged in it might have been fed from the public kitchen of the general establishment, and have been under the immediate inspection of the principal officers of the house of industry. It would likewise have rendered the establishment very interesting to those who visit it; which is an object of more real importance than can well be imagined by those who have not had occasion to know how much the approbation and applause of the public facilitate difficult enterprizes.

The means of uniting the rational amusement of society, with the furtherance of schemes calculated for the promotion of public good, is a subject highly deserving the attention of all who are engaged in public affairs.

CHAPTER. IX.

Of the means used for extending the influence of the institution
  for the poor at Munich, to other parts of Bavaria.
Of the progress which some of the improvements introduced at
Munich
  are making in other countries.

Though the institution of which I have undertaken to give an account, was confined to the city of Munich and its suburbs, yet measures were taken to extend its influence to all parts of the country. The attempt to put an end to mendicity in the capital, and to give employment to the Poor, having been completely successful, this event was formally announced to the public, in the newspapers; and other towns were called upon to follow the example. Not only a narrative in detail, was given of all the different measures pursued in this important undertaking, but every kind of information and assistance was afforded on the part of the institution at Munich, to all who might be disposed to engage in forming similar establishments in other parts of the country.

Copies of all the different lists, returns, certificates, etc. used in the management of the Poor, were given gratis to all, strangers as well as inhabitants of the country, who applied for them; and no information relative to the establishment, or to any of its details, was ever refused. The house of industry was open every day from morning till night to all visitors; and persons were appointed to accompany strangers in their tour through the different apartments, and to give the fullest information relative to the details, and even to all the secrets of the various manufactures carried on; and printed copies of the different tables, tickets, checks, etc. made use of in carrying on the current business of the house, were furnished to every one who asked for them; together with an account of the manner in which these were used, and of the other measures adopted to prevent frauds and peculation in the various branches of this extensive establishment.

As few manufactures in Bavaria are carried on to any extent; the more indigent of the inhabitants are, in general, so totally unacquainted with every kind of work in which the Poor could be most usefully employed, that that circumstance alone is a great obstacle to the general introduction throughout the country of the measures adopted in Munich for employing the Poor. To remove this difficulty, the different towns and communities who are desirous of forming establishments for giving employment to the Poor, are invited to send persons properly qualified to the house of industry at Munich, where they may be taught, gratis, spinning, in its various branches; knitting; sewing, etc. in order to qualify them to become instructors to the Poor on their return home. And even instructors already formed, and possessing all the requisite qualifications for such an office, are offered to be furnished by the house of industry in Munich to such communities as shall apply for them.

Another difficulty, apparently not less weighty than that just mentioned, but which is more easily and more effectually removed, is the embarrassment many of the smaller communities are likely to be under in procuring raw materials, and in selling to advantage the goods manufactured, or, (as is commonly the case,) IN PART ONLY MANUFACTURED, by the Poor. The yarn, for instance, which is spun by them in a country-town or village, far removed from any manufacture of cloth, may lie on hand a long time before it can be sold to advantage. To remedy this, the house of industry at Munich is ordered to furnish raw materials to such communities as shall apply for them, and receive in return the goods manufactured, at the full prices paid for the same articles in Munich. Not only these measures, and many others of a similar nature, are taken, to facilitate the introduction of industry among the Poor throughout the country; but every encouragement is held out to induce individuals to exert themselves in this laudable undertaking. Those communities which are the first to follow the example of the capital, are honourably mentioned in the news-papers; and such individuals as distinguished themselves by their zeal and activity upon those occasions, are praised and rewarded.

A worthy curate, (Mr. Lechner,) preacher in one of the churches in Munich, who, of his own accord, had taken upon himself to defend the measures adopted with regard to the Poor, and to recommend them in the most earnest manner from the pulpit, was sent for by the Elector, into his closet, and thanked for his exertions.

This transaction being immediately made known, (an account of it having been published in the news-papers,) tended not a little to engage the clergy in all parts of the country to exert themselves in support of the institution.

It is not my intention to insinuate that the clergy in Bavaria stood in need of any such motive to stimulate them to action in a cause so important to the happiness and well-being of mankind, and consequently so nearly connected with the sacred duties of their office; — on the contrary, I should be wanting in candour, as well as gratitude, were I not to embrace this opportunity of expressing publicity, the obligations I feel myself under to them for their support and assistance.

The number of excellent sermons which have been preached, in order to recommend the measures adopted by the government for making provision for the Poor, show how much this useful and respectable body of men have had it at heart to contribute to the success of this important measure; and their readiness to co-operate with me, (a Protestant,) upon all occasions where their assistance has been asked, not only does honour to the liberality of their sentiments, but calls for my personal acknowledgments, and particular thanks.

I shall conclude this Essay with an account of the progress which some of the improvements introduced at Munich are now making in other countries. During my late journey in Italy for the recovery of my health, I visited Verona; and becoming acquainted with the principal directors of two large and noble hospitals, la Pieta, and la Misericorde, in that city, the former containing about 350, and the latter near 500 Poor, I had frequent occasions to converse with them upon the subject of those establishments, and to give them an account of the arrangements that had been made in Munich. I likewise took the liberty of proposing some improvements, and particularly in regard to the arrangements for feeding these Poor; and in the management of the fires employed for cooking. Fire-wood, the only fuel used in that country, is extremely scarce and dear, and made a very heavy article in the expences of those institutions.

Though this scarcity of fuel, which had prevailed for ages in that part of Italy, had rendered it necessary to pay attention to the economy of fuel, and had occasioned some improvements to be made in the management of heat; yet I found, upon examining the kitchens of these two hospitals, and comparing the quantities of fuel consumed with the quantities of victuals cooked, that SEVERN-EIGHTHS of the fire-wood they were then consuming might be saved[15]. Having communicated the result of those enquiries to the directors of these two hospitals, and offered my service to alter the kitchens, and arrange them upon the principles of that in the house of industry at Munich, (which I described to them,) they accepted my offer, and the kitchens were rebuilt under my immediate direction; and have both succeeded, even beyond my most sanguine expectations. That of the hospital of la Pieta is the most complete kitchen I have ever built; and I would recommend it as a model, in preference to any I have ever seen. I shall give a more particular description of it, with plans and estimates, in my Essay on the Management of Heat.

During the time I was employed in building the new kitchen in the hospital of la Pieta, I had an opportunity of making myself acquainted with all the details of the clothing of the Poor belonging to that establishment; and I found that very great savings might be made in that article of expence. I made a proposal to the directors of that hospital, to furnish them with clothing for their Poor, ready made up, from the house of industry at Munich; and upon my return to Munich I sent them TWELVE complete suits of clothing of different sizes as a sample, and accompanied them with an estimate of the prices at which we could afford to deliver them at Verona.

The success of this little adventure has been very flattering, and has opened a very interesting channel for commerce, and for the encouragement of industry in Bavaria. This sample of clothing being approved, and, with all the expences of carriage added, being found to be near TWENTY PER CENT. cheaper than that formerly used, orders have been received from Italy by the house of industry at Munich, to a considerable amount, for clothing the Poor. In the beginning of September last, a few days before I left Munich to come to England, I had the pleasure to assist in packing up and sending off, over the Alps, by the Tyrol, SIX HUNDRED articles of

clothing of different kinds for the Poor of Verona; and hope soon to
see the Poor of Bavaria growing rich, by manufacturing clothing for
the Poor of Italy.

END OF THE FIRST ESSAY.

Footnotes to Essay I.

[1] This paper, as it could afterwards be made use of for making
cartridges, in fact cost nothing.

[2] A creutzer is 11/33 of an English penny.

[3] Particular local reasons, which it is not necessary here to explain, have hitherto prevented the establishment of military gardens
in these two garrison towns.

[4] The whole amount of this burden was not more than 30,000
florins, or about 2721L. sterling a year.

[5] Mons. Dallarmi.

[6] The annual amount of these various receipts may be seen in
the accounts published in the Appendix.

[7] Almost all the great law-givers, and founders of religions,
from the remotest antiquity, seem to have been aware of the influence of cleanliness upon the moral character of man; and have
strongly inculcated it. In many cases it has been interwoven with
the most solemn rites of public and private worship, and is so still
in many countries. The idea that the soul is defiled and depraved by
every thing UNCLEAN, or which defies the body, has certainly
prevailed in all ages; and has been particularly attended to by those
great benefactors of mankind, who, by the introduction of PEACE
and ORDER in society, have laboured successfully to promote the
happiness of their fellow-creatures. Order and disorder—peace and
war—health and sickness, cannot exist together; but COMFORT
and CONTENTMENT, and the inseparable companions of HAPPI-
NESS and VIRTUE, can only arise from order, peace, and health.

[8] Upon this occasion I must not forget to mention a curious circumstance, which contributed very much towards clearing the town

effectually of beggars. It being found that some of the most hardened of these vagabonds were attempting to return to their old practices, and that they found means to escape the patroles, by keeping a sharp look-out, and avoiding them, to hold them more effectually in check, the patroles sent out upon this service were ordered to go without arms. In consequence of this arrangement, the beggars being no longer able to distinguish who were in search of them, and who were not, saw a patrole in every soldier they met with in the streets, (and of these there were great numbers, Munich being a garrison town,) and from thenceforward they were kept in awe.

[9] Upon a new division of the town, when the suburbs were included, the number of subdivisions (abtheilungs) were augmented to twenty three.

[10] This was written in the summer of the year 1795.

[11] As these children were not shut up and confined like prisoners in the house of industry, but all lodged in the town, with their parents or friends, they had many opportunities to recreate themselves, and take exercise in the open air; not only on holidays, of which there are a very large number indeed kept in Bavaria; but also on working-days, in coming and going to and from the house of industry. Had not this been the case, a reasonable time would certainly have been allowed them for play and recreation. The cadets belonging to the Military Academy at Munich are allowed no less than THREE HOURS a day for exercise and relaxation, viz ONE HOUR immediately after dinner, which is devoted to music, and TWO HOURS, later in the afternoon, for walking in the country, or playing in the open fields near the town.

[12] Since the publication of the first edition of the Essay, the Author has received an account of the total destruction of the Military Work-house at Manheim. It was set on fire, and burnt to the ground, during the last siege of that city by the Austrian troops.

[13] Above 18,000 pounds sterling.

[14] The committee, at the head of the establishment, has been enabled to make this purchase, by legacies made to the institution. These legacies have been numerous, and are increasing every day;

which clearly shows, that the measures adopted with regard to the Poor have met with the approbation of the public.

[15] I found upon examining the famous kitchen of the great hospital at Florence, that the waste of fuel there is still greater.

CONTENTS of ESSAY II.

of the FUNDAMENTAL PRINCIPLES on which GENERAL ESTABLISHMENTS for the RELIEF of the POOR may be formed in all Countries.

ESSAY II.

CHAPTER. I.

Though the fundamental principles upon which the Establishment for the Poor at Munich is founded, are such as I can venture to recommend; and notwithstanding the fullest information relative to every part of that Establishment may, I believe, be collected from the account of it which is given in the foregoing Essay; yet, as this information is so dispersed in different parts of the work, and so blended with a variety of other particulars, that the reader would find some difficulty in bringing the whole into one view, and arranging it systematically in a complete whole; I shall endeavour briefly to resume the subject, and give the result of all my enquiries relative to it, in a more concise, methodical, and useful form: and as from the experience, I have had in providing for the wants of the Poor, and reclaiming the indolent and vicious to habits of useful industry, I may venture to consider myself authorised to speak with some degree of confidence upon the subject; instead of merely recapitulating what has been said of the Establishment for the Poor at Munich, (which would be at best but a tiresome repetition,) I shall now allow myself a greater range in these investigations, and shall give my opinions without restraint which may come under consideration. And though the system I shall propose, is founded upon the successful experiments made at Munich, as may be seen by comparing it with the details of that Establishment; yet, as a difference in the local circumstances under which an operation is performed, must necessarily require certain modifications of the plan, I shall endeavour to take due notice of every modification which may appear to me to be necessary[1].

Before I enter upon those details, it may be proper to take a more extensive survey of the subject, and investigate the general and fundamental Principles on which an Establishment for the Relief of the Poor, in every country, ought to be founded. At the same time I shall consider the difficulties which are generally understood to be inseparable from such an undertaking, and endeavour to show that they are by no means insurmountable.

That degree of poverty which involves in it the inability to procure the necessaries of life without the charitable assistance of the Public, is, doubtless, the heaviest of all misfortunes; as it not only brings along with it the greatest physical evils, pain,—and disease, but is attended by the most mortifying humiliation, and hopeless despondency. It is, moreover, an incurable evil; and is rather irritated than alleviated by the remedies commonly applied to remove it. The only alleviation, of which it is capable, must be derived from the kind and soothing attentions of the truly benevolent. This is the only balm which can sooth the anguish of a wounded heart, or allay the agitations of a mind irritated by disappointment, and rendered ferocious by despair.

And hence it evidently appears that no body of laws, however wisely framed, can, in any country, effectually provide for the relief of the Poor, without the voluntary assistance of individuals; for though taxes may be levied by authority of the laws for the support of the Poor, yet, those kind attentions which are so necessary in the management of the Poor, as well to reclaim the vicious, as to comfort and encourage the despondent—those demonstrations of concern which are always so great a consolation to persons in distress—cannot be COMMANDED BY FORCE. On the contrary, every attempt to use FORCE in such cases, seldom fails to produce consequences directly contrary to those intended[2].

But if the only effectual relief for the distress of the Poor, and the sovereign remedy for the numerous evils to society which arise from the prevalence of mendicity, indolence, poverty, and misery, among the lower classes of society, must be derived from the charitable and voluntary exertions of individuals;— as the assistance of the Public cannot be expected, unless the most unlimited confidence can be placed, not only in the wisdom of the measures proposed, but also, and MORE ESPECIALLY, in the UPRIGHTNESS, ZEAL, and PERFECT DISINTERESTEDNESS of the persons appointed to carry them into execution; it is evident that the first object to be attended to, in forming a plan of providing for the Poor, is to make such arrangements as will COMMAND THE CONFIDENCE OF THE PUBLIC, and fix it upon the most solid and durable foundation.

This can most certainly, and most effectually be done;
first by engaging persons of high rank and
  the most respectable character to place themselves
  at the head of the Establishment:
secondly, by joining, in the general administration of the
  affairs of the Establishment, a certain number of persons chosen
  from the middling class of society; reputable tradesmen, in easy
  circumstances;—heads of families;—and others of known integrity
  and of humane dispositions[3]:
thirdly, by engaging all those who are employed in the
  administration of the affairs of the Poor, to serve without fee
  or reward:
fourthly, by publishing, at stated periods, such particular and
  authentic accounts of all receipts and expenditures, that no
  doubt can possibly be entertained by the Public respecting the
  proper application of the monies destined for the relief of the
  Poor:
fifthly, by publishing an alphabetical list of all who receive
  alms; in which list should be inserted, not only the name of
  the person, his age; condition; and place of abode; but also
  the amount of the weekly assistance granted to him; in order
  that those who entertain any doubts respecting the manner in
  which the Poor are provided for, may have the opportunity of
  visiting them at their habitations, and making enquiry into
  their real situations:
and lastly, the confidence of the Public, and the
  continuance of their support, will most effectually be secured
  by a prompt and successful execution of the plan adopted.

There is scarcely a greater plague that can infest society, than swarms of beggars; and the inconveniencies to individuals arising from them are so generally, and so severely felt, that relief from so great an evil cannot fail to produce a powerful and lasting effect upon the minds of the Public, and to engage all ranks to unite in the support of measures as conducive to the comfort of individuals, as they are essential to the national honor and reputation. And even in countries where the Poor do not make a practice of begging, the knowledge of their sufferings must be painful to every benevolent mind; and there is no person, I would hope, so callous to the feelings of humanity, as not to rejoice most sincerely when effectual relief is afforded.

The greatest difficulty attending the introduction of any measure founded upon the voluntary support of the Public, for maintaining the Poor, and putting an end to mendicity, is an opinion generally entertained, that a very heavy expence would be indispensably necessary to carry into execution such an undertaking. But this difficulty may be speedily removed by showing, (which may easily be done,) that the execution of a well-arranged plan for providing for the Poor, and giving useful employment to the idle and indolent, so far from being expensive, must, in the end, be attended with a very considerable saving, not only to the Public collectively, but also to individuals.

Those who now extort their subsistence by begging and stealing, are, in fact, already maintained by the Public. But this is not all; they are maintained in a manner the most expensive and troublesome, to themselves and the Public, that can be conceived; and this may be said of all the Poor in general.

A poor person, who lives in poverty and misery, and merely from hand to mouth, has not the power of availing himself of any of those economical arrangements, in procuring the necessaries of life, which other, in more affluent circumstances, may employ; and which may be employed with peculiar advantage in a public Establishment. — Added to this, the greater part of the Poor, as well those who make a profession of begging, as other who do not, might be usefully employed in various kinds of labour; and supposing them, one with another, to be capable of earning ONLY HALF as much as is necessary to their subsistence, this would reduce the present expence to the Public for their maintenance at least one half; and this half might be reduced still much lower, by a proper attention to order and economy in providing for their subsistence.

Were the inhabitants of a large town where mendicity is prevalent, to subscribe only half the sums annually, which are extorted from them by beggars, I am confident it would be quite sufficient, with a proper arrangement, for the comfortable support of the Poor of all denominations.

Not only those who were formerly common street-beggars, but all others, without exception, who receive alms, in the city of Munich and its suburbs, amounting at this time to more than 1800 persons, are supported almost entirely by voluntary subscriptions from the inhabitants; and I have been assured by numbers of the most opulent and respectable citizens, that the sums annually extorted from them formerly by beggars alone, exclusive of private charities, amounted to more than three times the sums now given by them to the support of the new institution. I insist the more upon this point, as I know that the great expence which has been supposed to be indispensably necessary to carry into execution any scheme for effectually providing for the Poor, and putting an end to mendicity, has deterred many well-disposed persons from engaging in so useful an enterprise. I have only to add my most earnest wishes, that what I have said and done, may remove every doubt, and re-animate the zeal of the Public, in a cause in which the dearest interests of humanity are so nearly concerned.

In almost every public undertaking, which is to be carried into effect by the united voluntary exertions of individuals, without the interference of government, there is a degree of awkwardness in bringing forward the business, which it is difficult to avoid, and which is frequently not a little embarrassing. This will doubtless be felt by those who engage in forming and executing schemes for providing for the Poor by private subscription; they should not, however, suffer themselves to be discouraged by a difficulty which may so easily be surmounted.

In the introduction of every scheme for forming an Establishment for the Poor, whether it be proposed to defray the expense by voluntary subscriptions, or by a tax levied for the purpose, it will be proper for the authors or promoters of the measure to address the Public upon the subject; to inform them of the nature of the measures proposed;— of their tendency to promote the public welfare, and to point out the various ways in which individuals may give their assistance to render the scheme successful.

There are few cities in Europe, I believe, in which the state of the Poor would justify such an address as that which was published at Munich upon taking up the beggars in that town; but something of the kind; with such alterations as local circumstances may require, I am persuaded, would in most cases produce good effects. With regard to the assistance that might be be given by individuals to carry into effect a scheme for providing for the Poor, though measures for that purpose may, and ought to be so taken, that the Public would have little or no trouble in their execution, yet there are many things which individuals must be instructed cautiously to avoid; otherwise the enterprise will be extremely difficult, it not impracticable; and, above all things, they must be warned against giving alms to beggars.

Though nothing would be more unjust and tyrannical, than to prevent the generous and humane from contributing to the relief of the Poor and necessitous, yet, as giving alms to beggars tends so directly and so powerfully to encourage idleness and immorality, to discourage the industrious Poor, and perpetuate mendicity, with all its attendant evils, too much pains cannot be taken to guard the Public against a practice so fatal in its consequences to society.

All who are desirous of contributing to the relief of the Poor, should be invited to send their charitable donations to be distributed by those who, being at the head of a public Institution established for taking care of the Poor, must be supposed best acquainted with their wants. Or, if individuals should prefer distributing their own charities, they ought at least to take the trouble to enquire after fit objects; and to apply their donations in such a manner as not to counteract the measures of a public and useful Establishment.

But, before I enter farther into these details, it will be necessary to determine the proper extent and limits of an Establishment for the

Poor; and show how a town or city ought to be divided in districts, in order to facilitate the purposes of such an institution.

CHAPTER. II. Of the Extent of an Establishment for the Poor. Of the Division of a Town or City into Districts. Of the Manner of carrying on the Business of a public Establishment for the Poor. Of the Necessity of numbering all the Houses in a Town where an Establishment for the Poor is formed.

However large a city may be, in which an Establishment for the Poor is to be formed, I am clearly of opinion, that there should be but ONE ESTABLISHMENT;—with ONE committee for the general management of all its affairs;—and ONE treasurer. This unity appears essentially necessary, not only because, when all the parts tend to one common centre, and act in union to the same end, under one direction, they are less liable to be impeded in their operations, or disordered by collision;—but also on account of THE VERY UNEQUAL DISTRIBUTION OF WEALTH, as well as of misery and poverty, in the different districts of the same town. Some parishes in great cities have comparatively few Poor, while others, perhaps less opulent, are overburthened with them; and there seems to be no good reason why a house-keeper in any town should be called upon to pay more or less for the support of the Poor, because he happens to live on one side of a street or the other. Added to this, there are certain districts in most great towns where poverty and misery seem to have fixed their head-quarters, and where it would be IMPOSSIBLE for the inhabitants to support the expence of maintaining their Poor. Where that is the case, as measures for preventing mendicity in every town must be general, in order to their being successful, the enterprise, FROM THAT CIRCUMSTANCE ALONE, would be rendered impracticable, were the assistance of the more opulent districts to be refused.

There is a district, for instance, belonging to Munich, (the Au,) a very large parish, which may be called the St. Giles's of that city, where the alms annually received are TWENTY TIMES as much as the whole district contributes to the funds of the public Institution for the Poor.—The inhabitants of the other parishes, however, have never considered it a hardship to them, that the Poor of the Au should be admitted to share the public bounty, in common with the Poor of the other parishes.

Every town must be divided, according to its extent, into a greater or less number of districts, or subdivisions; and each of these must have a committee of inspection, or rather a commissary, with assistants, who must be entrusted with the superintendance and management of all affairs relative to the relief and support of the Poor within its limits.

In very large cities, as the details of a general Establishment for the Poor would be very numerous and extensive, it would probably facilitate the management of the affairs of the Establishment, if, beside the smallest subdivisions or districts, there could be formed other larger divisions, composed of a certain number of districts, and put under the direction of particular committees.

The most natural, and perhaps the most convenient method of dividing a large city or town, for the purpose of introducing a general Establishment for the Poor, would be, to form of the parishes the primary divisions; and to divide each parish into so many subdivisions, or districts, as that each district may consist of from 3000 to 4000 inhabitants. Though the immediate inspection and general superintendance of the affairs of each parish were to be left to its own particular committee, yet the supreme committee at the head of the general Institution should not only exercise a controlling power over the parochial committees, but these last should not be empowered to levy money upon the parishioners, by setting on foot voluntary subscriptions, or otherwise; or to dispose of any sums belonging to the general Institution, except in cases of urgent necessity;—nor should they be permitted to introduce any new arrangements with respect to the management of the Poor, without the approbation and consent of the supreme committee: the most perfect uniformity in the mode of treating the Poor, and transacting all public business relative to the Institution, being indispensably necessary to secure success to the undertaking, and fix the Establishment upon a firm and durable foundation.

For the same reasons, all monies collected in the parishes should not be received and disposed of by their particular committees, but ought to be paid into the public treasury of the Institution, and carried to the general account of receipts;—and, in like manner, the sums necessary for the support of the Poor in each parish should be furnished from the general treasury, on the orders of the supreme committee.

With regard to the applications of individuals in distress for assistance, all such applications ought to be made through the commissary of the district to the parochial committee;—and where the necessity is not urgent, and particularly where permanent assistance is required, the demand should be referred by the parochial committee to the supreme committee, for their decision. In cases of urgent necessity, the parochial committees, and even the commissaries of districts, should be authorized to administer relief, ex officio, and without delay; for which purpose they should be furnished with certain sums in advance, to be afterwards accounted for by them.

That the supreme committee may be exactly informed of the real state of those in distress who apply for relief, every petition, forwarded by a parochial committee, or by a commissary of a district, where there are no parochial committees, should be accompanied with an exact and detailed account of the circumstances of the petitioner, signed by the commissary of the district to which he belongs, together with the amount of the weekly sum, or other relief, which such commissary may deem necessary for the support of the petitioner.

To save the commissaries of districts the trouble of writing the descriptions of the Poor who apply for assistance, printed forms, similar to that which may been seen in the Appendix, No. V. may be furnished to them;—and other printed forms, of a like nature, may be introduced with great advantage in many other cafes in the management of the Poor.

With regard to the manner in which the supreme and parochial committees should be formed;— however they may be composed, it will be indispensably requisite, for the preservation of order and harmony in all the different parts of the Establishment, that one member at least of each parochial committee be present, and have a seat, and voice, as a member of the supreme committee. And, that all the members of each parochial committee may be equally well informed with regard to the general affairs of the Establishment, it may perhaps be proper that those members attended the meetings of the supreme committee in rotation.

For similar reasons it may be proper to invite the commissaries of districts to be present in rotation at the meetings of the committees of their respective parishes, where there are parochial committees established, or otherwise, at the meetings of the supreme committee[4].

It is, however, only in very large cities that I would recommend the forming parochial committees. In all towns where the inhabitants do not amount to more than 100,000 souls, I am clearly of opinion that it would be best merely to divide the town into districts, without regard to the limits of parishes; and to direct all the affairs of the institution by one simple committee. This mode was adopted at Munich, and found to be easy in practice, and successful; and it is not without some degree of diffidence, I own, that I have ventured to propose a deviation from a plan, which has not yet been justified by experience.

But however a town may be divided into districts, it will be absolutely necessary that ALL the houses be regularly numbered, and an accurate list made out of all the persons who inhabit them. The propriety of this measure is too apparent to require any particular explanation. It is one of the very first steps that ought to be taken in carrying into execution any plan for forming an Establishment for the Poor; it being as necessary to know the names and places of abode of those, who, by voluntary subscription, or otherwise, assist in relieving the Poor, as to be acquainted with the dwellings of the objects themselves; and this measure is as indispensable necessary when an institution for the Poor is formed in a small country-town or village, as when it is formed in the largest capital.

In many cases, it is probable, the established laws of the country in which an institution for the Poor may be formed, and certain usages, the influence of which may perhaps be still more powerful than the laws, may render modifications necessary, which it is utterly impossible for me to foresee; still the great fundamental principles upon which every sensible plan for such an Establishment must be founded, appear to me to be certain and immutable; and when rightly understood, there can be no great difficulty in accommodating the plan to all those particular circumstances under which it may be carried into execution, without making any essential alteration.

CHAPTER. III.

Whatever the number of districts into which a city is divided, may be, or the number of committees employed in the management of a public Establishment for the relief of the Poor, it is indispensably necessary that all individuals who are employed in the undertaking be persons of known integrity;—for courage is not more necessary in the character of a general, than unshaken integrity in the character of a governor of a public charity. I insist the more upon this point as the whole scheme is founded upon the voluntary assistance of individuals, and therefore to ensure its success the most unlimited confidence of the public must be reposed in those who are to carry it into execution; besides, I may add, that the manner in which the funds of the various public Establishments for the relief of the Poor already instituted have been commonly been administered in most countries, does not tend to render superfluous the precautions I propose for securing the confidence of the public.

The preceding observations respecting the importance of employing none but persons of known integrity at the head of an institution for the relief of the Poor, relates chiefly to the necessity of encouraging people in affluent circumstances, and the public at large, to unite in the support of such an Establishment.—There is also another reason, perhaps equally important, which renders it expedient to employ persons of the most respectable character in the details of an institution of public charity,—the good effects such a choice must have upon the minds and morals of the Poor.

Persons who are reduced to indigent circumstances, and become objects of public charity, come under the direction of those who are appointed to take care of them with minds weakened by adversity, and soured by disappointment; and finding themselves separated from the rest of mankind, and cut off from all hope of seeing better days, they naturally grow peevish, and discontented, suspicious of those set over them, and of one another; and the kindest treatment, and most careful attention to every circumstance that can render their situation supportable, are therefore required, to prevent their being very unhappy. And nothing surely can contribute more powerfully to soothe the minds of persons in such unfortunate and hopeless circumstances, than to find themselves under the care and protection of persons of gentle manners;—humane dispositions;—and known probity and integrity; such as even THEY,—with all their suspicions about them, may venture to love and respect,

Whoever has taken the pains to investigate the nature of the human mind, and examine attentively those circumstances upon which human happiness depends, must know how necessary it is to happiness, that the mind should have some object upon which to place its more tender affections—something to love,—to cherish, —to esteem,—to respect,—and to venerate; and these resources are never so necessary as in the hour of adversity and discouragement, where no ray of hope is left to cheer the prospect, and stimulate to fresh exertion.

The lot of the Poor, particularly of those who, from easy circumstances and a reputable station in society, are reduced by misfortunes, or oppression, to become a burthen on the Public, is truly deplorable, after all that can be done for them:— and were we seriously to consider their situation, I am sure we should think that we could never do too much to alleviate their sufferings, and soothe the anguish of wounds which can never be healed.

For the common misfortunes of life, HOPE is a sovereign remedy. But what remedy can be applied to evils, which involve even the loss of hope itself? and what can those have to hope, who are separated and cut off from society, and for ever excluded from all share in the affairs of men? To them, honours;—distinctions; —praise;— and even property itself;—all those objects of laudable ambition which so powerfully excite the activity of man in civil society, and contribute so essentially to happiness, by filling the mind with pleasing prospects of future enjoyments, are but empty names; or rather, they are subjects of never-ceasing regret and discontent.

That gloom must indeed be dreadful, which overspreads the mind, when HOPE, that bright luminary of the soul, which enlightens and cheers it, and excites and calls forth into action all its best faculties, has disappeared!

There are many, it is true, who, from their indolence or extravagance, or other vicious habits, fall into poverty and distress, and become a burthen on the public, who are so vile and degenerate as not to feel the wretchedness of their situation. But these are miserable objects, which the truly benevolent will regard with an eye of peculiar compassion;—they must be very unhappy, for they are very vicious; and nothing should be omitted, that can tend to reclaim them;—but nothing will tend so powerfully to reform them, as kind usage from the hands of persons they must learn to love and to respect at the same time.

If I am too prolix upon this head, I am sorry for it. It is a strong conviction of the great importance of the subject, which carries me away, and makes me, perhaps, tiresome, where I would wish most to avoid it. The care of the Poor, however, I must consider as a matter of very serious importance. It appears to me to be one of the most sacred duties imposed upon men in a state of civil society;— one of those duties imposed immediately by the hand of God himself, and of which the neglect never goes unpunished.

What I have said respecting the necessary qualifications of those employed in taking care of the Poor, I hope will not deter well-disposed persons, who are willing to assist in so useful an undertaking, from coming forward with propositions for the institution of public Establishments for that purpose; or from offering themselves candidates for employments in the management of such Establishments. The qualifications pointed out, integrity, and a gentle and humane disposition,—honesty, and a good heart;— are such as any one may boldly lay claim to, without fear of being taxed with vanity or ostentation. — And if individuals in private stations, on any occasion are called upon to lay aside their bashfulness and modest dissidence, and come forward into public view, it must surely be, when by their exertions they can essentially contribute to promote measures which are calculated to increase the happiness and prosperity of society.

It is a vulgar saying, that, what is everybody's business, is nobody's business; and it is very certain that many schemes, evidently intended for the public good, have been neglected, merely because nobody could be prevailed on to stand forward and be the first to adopt them. This doubtless has been the case in regard to many judicious and well arranged proposals for providing for the Poor; and will probably be so again. I shall endeavour, however, to show, that though in undertakings in which the general welfare of society is concerned, persons of all ranks and conditions are called upon to give them their support, yet, in the INTRODUCTION of such measures as are here recommended, — a scheme of providing for the Poor, — there are many who, by their rank and peculiar situations, are clearly pointed out as the most proper to take up the business at its commencement, and bring it forward to maturity; as well as to take an active part in the direction and management of such an institution after it has been established: and it appears to me, that the nature and the end of the undertaking evidently point out the persons who are more particularly called upon to set an example on such an occasion.

If the care of the Poor be an object of great national importance, — if it be inseparably connected with the peace and tranquillity of society, and with the glory and prosperity of the state; — if the advantages which individuals share in the public welfare are in proportion to the capital they have at stake in this great national fund — that is to say, in proportion to their rank, property, and connexions, or general influence; — as it is just that every one should contribute in proportion to the advantages he receives; it is evident who ought to be the first to come forward upon such an occasion.

But it is not merely on account of the superior interest they have in the public welfare, that persons of high rank and great property, and such as occupy places of importance in the government, are bound to support measures calculated to relieve the distresses of the Poor;—there is still another circumstance which renders it indispensably necessary that they should take an active part in such measures, and that is, the influence which their example must have upon others.

It is impossible to prevent the bulk of mankind from being swayed by the example of those to whom they are taught to look up as their superiors; it behoves, therefore, all who enjoy such high privileges, to employ all the influence which their rank and fortune give them, to promote the public good. And this may justly be considered as a duty of a peculiar kind;—a PERSONAL service attached to the station they hold in society, and which cannot be commuted.

But if the obligations which persons of rank and property are under, to support measures designed for the relief of the Poor, are so binding, how much more so must they be upon those who have taken upon themselves the sacred office of public teachers of virtue and morality;—the Ministers of a most holy religion;— a religion whose first precepts inculcate charity and universal benevolence, and whose great object is, unquestionably, the peace, order, and happiness of society.

If there be any whose peculiar province it is to seek for objects in distress and want, and administer to them relief;—if there be any who are bound by the indispensable duties of their profession to encourage by every means in their power, and more especially by EXAMPLE, the general practice of charity; it is, doubtless, the Ministers of the gospel. And such is their influence in society, arising from the nature of their office, that their example is a matter of VERY SERIOUS IMPORTANCE.

Little persuasion, I should hope, would be necessary to induce the clergy, in any country, to give their cordial and active assistance in relieving the distresses of the Poor, and providing for their comfort and happiness, by introducing order and useful industry among them.

Another class of men, who from the station they hold in society, and their knowledge of the laws of the country, may be highly useful in carrying into effect such an undertaking, are the civil magistrates; and, however a committee for the government and direction of an Establishment for the Poor may in other respects be composed, I am clearly of opinion, that the Chief Magistrate of the town, or city, where such an Establishment is formed, ought always to be one of its members. The Clergyman of the place who is highest in rank or dignity ought, likewise, to be another; and if he be a Bishop, or Archbishop, his assistance is the more indispensable.

But as persons who hold offices of great trust and importance in the church, as well as under the civil government, may be so much engaged in the duties of their stations, as not to have sufficient leisure to attend to other matters; it may be necessary, when such distinguished persons lend their assistance in the management of an Establishment for the relief of the Poor, that each of them be permitted to bring with them a person of his own choice into the committee, to assist him in the business. The Bishop, for instance, may bring his chaplain;—the Magistrate, his clerk;—the Nobleman, or private gentleman, his son, or friend, etc. But in small towns, of two or three parishes, and particularly in country-towns and villages, which do not consist of more than one or two parishes, as the details in the management of the affairs of the Poor in such communities cannot be extensive, the members of the committee may manage the business without assistants. And indeed in all cases, even in great cities, when a general Establishment for the Poor is formed upon a good plan, the details of the executive and more laborious parts of the management of it will be so divided among the commissaries of the districts, that the members of the supreme committee will have little more to do than just hold the reins, and direct the movement of the machine. Care must however be taken to preserve the most perfect uniformity in the motions of all its parts, otherwise confusion must ensue; hence the necessity of directing the whole from one center.

As the inspection of the Poor;—the care of them when they are sick;—the distribution of the sums granted in alms for their support;—the furnishing them with clothes;—and the collection of the voluntary subscriptions of the inhabitants,—will be performed by the commissaries of the districts, and their assistants;—and as all the details relative to giving employment to the Poor, and feeding them, may be managed by particular subordinate committees, appointed for those purposes, the current business of the supreme committee will amount to little more than the exercise of a general superintendance.

This committee, it is true, must determine upon all demands from the Poor who apply for assistance; but as every such demand will be accompanied with the most particular account of the circumstances of the petitioner, and the nature and amount of assistance necessary to his relief, certified by the commissary of the district in which the petitioner resides, — and also by the parochial committee, where such are established, — the matter will be so prepared and digested, that the members of the supreme committee will have very little trouble to decide on the merits of the case, and the assistance to be granted.

This assistance will consist — in a certain sum to be given WEEKLY in alms to the petitioner, by the commissary of the district, out of the funds of the Institution; — in an allowance of bread only; — in a present of certain articles of clothing, which will be specified; — or, perhaps, merely in an order for being furnished with wood, clothing, or fuel, from the public kitchens or magazines of the Establishment, AT THE PRIME COST of those articles, AS AN ASSISTANCE to the petitioner, and to prevent the NECESSITY OF HIS BECOMING A BURTHEN ON THE PUBLIC.

The manner last mentioned of assisting the Poor, — that of furnishing them with the necessaries of life at lower prices than those at which they are sold in the public markets, is a matter of such importance, that I shall take occasion to treat of it more fully hereafter.

With respect to the petitions presented to the committee; — whatever be the assistance demanded, the petition received ought to be accompanied by a duplicate; to the end that, the decision of the committee being entered upon the duplicate, as well as upon the original, and the duplicate sent back to the commissary of the district, the business may be finished with the least trouble possible; and even without the necessary of any more formal order relative to the matter being given by the committee.

I have already mentioned the great utility of PRINTED FORMS, for petitions, returns, etc. in carrying on the business of an Establishment for the Poor, and I would again most earnestly recommend the general use of them. Those who have not had experience in such matters, can have no idea how much they contribute to preserve order, and facilitate and expedite business. To the general introduction of them in the management of the affairs of the Institution for the Poor at Munich, I attribute, more than to any thing else, the perfect order which has continued to reign throughout every part of that extensive Establishment, from its first existence to the present moment.

In carrying on the business of that Establishment, printed forms or blanks are used, not only for petitions;—returns;—lists of the Poor;— descriptions of the Poor;—lists of the inhabitants; —lists of subscribers to the support of the Poor;—orders upon the banker or treasurer of the Institution;—but also for the reports of the monthly collections made by the commissaries of districts;—the accounts sent in by the commissaries, of the extraordinary expences incurred in affording assistance to those who stand in need of immediate relief;—the banker's receipts; —and even the books in which are kept the accounts of the receipts and expenditures of the Establishment.

In regard to the proper forms for these blanks; as they must depend, in a great measure, upon local circumstances, no general directions can be given, other than, in all cases, the shortest forms that can be drawn up, consistent with perspicuity, are recommended; and that the subject-matter of each particular or single return, may be so disposed as to be easily transferred to such general tables, or general accounts, as the nature of the return and other circumstances may require. Care should likewise be taken to make them of such a form, SHAPE and dimension, that they may be regularly folded up, and docketed, in order to their being preserved among the public records of the Institution.

CHAPTER. IV.

Of the Necessity of effectual Measures for introducing a Spirit
  of Industry among the Poor in forming an Establishment for
  their Relief and Support.
Of the Means which may be used for that Purpose; and for setting
  on foot a Scheme for forming an Establishment for feeding the
  Poor.

An object of the very first importance in forming an Establishment for the relief and support of the Poor, is to take effectual measures for introducing a spirit of industry among them; for it is most certain, that all sums of money, or other assistance, given to the Poor in alms, which do not tend to make them industrious, never can fail to have a contrary tendency, and to operate as an encouragement to idleness and immorality.

And as the merit of an action is to be determined by the good it produces, the charity of a nation ought not to be estimated by the millions which are paid in Poor's taxes, but by the pains which are taken to see that the sums raised are properly applied.

As the providing useful employment for the Poor, and rendering them industrious, is, and ever has been, a great DESIDERATUM in political economy, it may be proper to enlarge a little here, upon that interesting subject.

The great mistake committed in most of the attempts which have been made to introduce a spirit of industry, where habits of idleness have prevailed, has been the too frequent and improper use of coercive measures, by which the persons to be reclaimed have commonly been offended and thoroughly disgusted at the very out-set. — Force will not do it. — Address, not force, must be used on those occasions.

The children in the house of industry at Munich, who, being placed upon elevated seats round the halls where other children worked, were made to be idle spectators of that amusing scene, cried most bitterly when their request to be permitted to descend from their places, and mix in that busy crowd, was refused; — but they would, most probably, have cried still more, had they been taken abruptly from their play and FORCED to work.

"Men are but children of a larger growth;" — and those who undertake to direct them, ought ever to bear in mind that important truth.

That impatience of control, and jealousy and obstinate perseverance in maintaining the rights of personal liberty and independence, which so strongly mark the human character in all the stages of life, must be managed with great caution and address, by those who are desirous of doing good; — or, indeed, of doing any thing effectually with mankind.

It has often been said, that the Poor are vicious and profligate, and that THEREFORE nothing but force will answer to make them obedient, and keep them in order; — but, I should say, that BECAUSE the Poor are vicious and profligate, it is so much the more necessary to avoid the appearance of force in the management of them, to prevent their becoming rebellious and incorrigible.

Those who are employed to take up and tame the wild horses belonging to the Elector Palatine, which are bred in the forest near Dusseldorf, never use force in reclaiming that noble animal, and making him docile and obedient. They begin with making a great circuit, in order to approach him; and rather decoy than force him into the situation in which they wish to bring him, and ever afterwards treat him with the greatest kindness; it having been found by experience, that ill-usage seldom fails to make him "a man-hater," untameable, and incorrigibly vicious. — It may, perhaps, be thought fanciful and trifling, but the fact really is, that an attention to the means used by these people to gain the confidence of those animals, and teach them to like their keepers, their stables, and their mangers, suggested to me many ideas which I afterwards put in execution with great success, in reclaiming those abandoned and ferocious animals in human shape, which I undertook to tame and render gentle and docile.

It is however necessary in every attempt to introduce a spirit of order and industry among the idle and profligate, not merely to avoid all harsh and offensive treatment, which, as has already been observed, could only serve to irritate them and render them still more vicious and obstinate, but it is also indispensably necessary to do every thing that can be devised to encourage and reward every symptom of reformation.

It will likewise be necessary sometimes to punish the obstinate; but recourse should never be had to punishments till GOOD USAGE has first been fairly tried and found to be ineffectual. The delinquent must be made to see that he has deserved the punishment, and when it is inflicted, care should be taken to make him feel it. But in order that the punishment may have the effects intended, and not serve to irritate the person punished, and excite personal hatred and revenge, instead of disposing the mind to serious reflection, it must be administered in the most solemn and most DISPASSIONATE manner; and it must be continued no longer than till the FIRST DAWN of reformation appears.

How much prudence and caution are necessary in dispensing rewards and punishments;—and yet—how little attention is in general paid to those important transactions!

REWARDS and PUNISHMENTS are the only means by which mankind can be controlled and directed; and yet, how often do we see them dispensed in the most careless—most imprudent—and most improper manner!—how often are they confounded!—how often misapplied!— and how often do we see them made the instruments of gratifying the most sordid private passions!

To the improper use of them may be attributed all the disorders of civil society.—To the improper or careless use of them may, most unquestionably, be attributed the prevalence of poverty, misery, and mendicity in most countries, and particularly in Great Britain, where the healthfulness and mildness of the climate—the fertility of the soil—the abundance of fuel—the numerous and flourishing manufactures—the extensive commerce— and the millions of acres of waste lands which still remain to be cultivated, furnish the means of giving useful employment to all its inhabitants, and even to a much more numerous population.

But if instead of encouraging the laudable exertions of useful industry, and assisting and relieving the unfortunate and the infirm— (the only real objects of charity,)—the means designed for those purposes are so misapplied as to operate as rewards to idleness and immorality, the greater the sums are which are levied on the rich for the relief of the poor, the more numerous will that class become, and the greater will be their profligacy, their insolence, and their shameless and clamorous importunity.

There is, it cannot be denied, in man, a natural propensity to sloth and indolence; and though habits of industry,—like all habits,— may render those exertions easy and pleasant which at first are painful and irksome, yet no person, in any situation, ever chose labour merely for its own sake. It is always the apprehension of some greater evil,—or the hope of some enjoyment, by which mankind are compelled or allured, when they take to industrious pursuits.

In the rude state of savage nature the wants of men are few, and these may all be easily supplied without the commission of any crime; consequently industry, under such circumstances, is not necessary, nor can indolence be justly considered as a vice; but in a state of civil society, where population is great, and the means of subsistence not to be had without labour, or without defrauding others of the fruits of their industry, idleness becomes a crime of the most fatal tendency, and consequently of the most heinous nature; and every means should be used to discountenance, punish, and prevent it.

And we see that Providence, ever attentive to provide remedies for the disorders which the progress of society occasions in the world, has provided for idleness—as soon as the condition of society renders it a vice, but not before—a punishment every way suited to its nature, and calculated to prevent its prevalency and pernicious consequences:—This is WANT,—and a most efficacious remedy it is for the evil,—when the WISDOM OF MAN does not interfere to counteract it, and prevent its salutary effects.

But reserving the father investigation of this part of my subject — that respecting the means to be used for encouraging industry— to some future opportunity, I shall now endeavour to show, in a few words, how, under the most unfavourable circumstances, an arrangement for putting an end to mendicity, and introducing a spirit of industry among the Poor, might be introduced and carried into execution.

If I am obliged to take a great circuit, in order to arrive at my object, it must be remembered, that where a vast weight is to be raised by human means, a variety of machinery must necessarily be provided; and that it is only by bringing all the different powers employed to act together to the same end, that the purpose in view can be attained. It will likewise be remembered, that as no mechanical power can be made to act without a force be applied to it sufficient to overcome the resistance, not only of the vis inertia, but also of friction, so no moral agent can be brought to act to any given end without sufficient motives; that is to say, without such motives as THE PERSON WHO IS TO ACT may deem sufficient, not only to decide his opinion, but also to OVERCOME HIS INDOLENCE.

The object proposed,—the relief of the Poor, and the providing for their future comfort and happiness, by introducing among them a spirit of order and industry, is such as cannot fail to meet with the approbation of every well-disposed person.—But I will suppose, that a bare conviction of the UTILITY of the measure is not sufficient alone to overcome the indolence of the Public, and induce them to engage ACTIVELY in the undertaking;—yet as people are at all times, and in all situations, ready enough to do what they FEEL to be their interest, if, in bringing forward a scheme of public utility, the proper means be used to render it so interesting as to awaken the CURIOSITY, and fix the attention of the Public, no doubts can be entertained of the possibility of carrying it into effect.

In arranging such a plan, and laying it before the Public, no small degree of knowledge of mankind, and particularly of the various means of acting on them, which are peculiarly adapted to the different stages of civilization, or rather of the political refinement and corruption of society, would, in most cases, be indispensably necessary; but with that knowledge, and a good share of zeal, address, prudence, and perseverance, there are few schemes, in which an honest man would wish to be concerned, that might not be carried into execution in any country.

In such a city as London, where there is great wealth;—public spirit;—enterprize;—and zeal for improvement; little more, I flatter myself, would be necessary to engage all ranks to unite in carrying into effect such a scheme, than to show its public utility; and, above all, to prove that there IS NO JOB at the bottom of it.

It would, however, be advisable, in submitting to the Public, Proposals for forming such an Establishment, to show that those who are invited to assist in carrying it into execution, would not only derive from it much pleasure and satisfaction, but also many real advantages; for too much pains can never be taken to interest the Public individually, and directly, in the success of measures tending to promote the general good of society.

The following Proposals, which I will suppose to be made by some person of known and respectable character, who has courage enough to engage in so arduous an undertaking, will show my ideas upon this subject in the clearest manner. — Whether they are well founded, must be left to the reader to determine. — As to myself, I am so much persuaded that the scheme here proposed, by way of example, and merely for illustration, might be executed, that, had I time for the undertaking, (which I have not,) I should not hesitate to engage in it.

PROPOSALS for forming by private subscription, an ESTABLISHMENT for feeding the Poor, and giving them useful Employment;

And also for furnishing Food at a cheap Rate to others who may stand in need of such Assistance. Connected with an INSTITUTION for introducing, and bringing forward into general Use, new Inventions and Improvements, particularly such as relate to the Management of Heat and the saving of Fuel; and to various other mechanical Contrivances by which DOMESTIC COMFORT and ECONOMY may be promoted. Submitted to the Public, By A. B.

The Author of these Proposals declares solemnly, in the face of the whole world, that he has no interested view whatever in making these Proposals; but is actuated merely and simply by a desire to do good, and promote the happiness and prosperity of society, and the honour and reputation of his country. — That he never will demand, accept, or receive any pay or other recompence or reward of any kind whatever from any person or persons, for his services or trouble, in carrying into execution the proposed scheme, or any part thereof, or for anything he may do or perform in future relating to it, or to any of its details or concerns.

And, moreover, that he never will avail himself of any opportunities that may offer in the execution of the plan proposed, for deriving profit, emolument, or advantage of any kind, either for himself, his friends, or connections;—but that, on the contrary, he will take upon himself to be personally responsible to the Public, and more immediately to the Subscribers to this Undertaking, that NO PERSON shall FIND MEANS to make a job of the proposed Establishment, or of any of the details of its execution, or of its management, as long as the Author of these Proposals remains charged with its direction.

With respect to the particular objects and extent of the proposed Establishment, these may be seen by the account which is given of them at the head of these Proposals; and as to their utility, there can be no doubts. They certainly must tend very powerfully to promote the comfort, happiness, and prosperity of society, and will do honour to the nation, as well as to those individuals who may contribute to carry them into execution.

With the regard to the possibility of carrying into effect the proposed scheme;—the facility with which this may be done, will be evident when the method of doing it, which will now be pointed out, is duly considered.

As soon as a sum shall be subscribed sufficient for the purposes intended, the Author of these Proposals will, by letters, request a meeting of the TWENTY-FIVE persons who shall stand highest on the list of subscribers, for the purpose of examining the subscription-lists, and of appointing, by ballot, a committee, composed of five persons, skilled in the details of building, and in accounts, to collect the subscriptions, and to superintend the execution of the plan.—This committee, which will be chosen from among the subscribers at large, will be authorised and directed, to examine all the works that will be necessary in forming the Establishment, and see that they are properly performed, and at reasonable prices;—to examine and approve of all contracts for work, or for materials;—to examine and check all accounts of expenditures of every kind, in the execution of the plan;—and to give orders for all payments.

The general arrangement of the Establishment, and of all its details, will be left to the Author of these Proposals; who will be responsible for their success. — He engages, however, in the prosecution of this business, to adhere faithfully to the plan here proposed, and never to depart from it on any pretence whatever.

With regard to the choice of a spot for erecting this Establishment, a place will be chosen within the limits of the town, and in a convenient and central a situation as possible, where ground enough for the purpose is to be had at a reasonable price[5]. — The agreement for the purchase, or hire of this ground, and of the buildings, if there be any on it, will, like all other bargains and contracts, be submitted to the committee for their approbation and ratification.

The order in which it is proposed to carry into execution the different parts of the scheme is as follows: — First, to establish a public kitchen for furnishing Food to such poor persons as shall be recommended by the subscribers for such assistance.

This Food will be of four different sorts, namely, No. I. A nourishing soup composed of barley — pease — potatoes, and bread; seasoned with salt, pepper, and fine herbs. — The portion of this soup, one pint and a quarter, weighing about twenty ounces, will cost ONE PENNY.

No. II. A rich pease-soup, well seasoned; — with fried bread; — the portion (twenty ounces) at TWO PENCE.

No. III. A rich and nourishing soup, or barley, pease, and potatoes, properly seasoned; — with fried bread; and two ounces of boiled bacon, cut fine and put into it. — The portion (20 ounces) at FOUR PENCE.

No. IV. A good soup; with boiled meat and potatoes or cabbages, or other vegetables; with 1/4 lb. of good rye bread, the portion at SIX PENCE.

Adjoining to the kitchen, four spacious eating-rooms will be fitted up, in each of which one only of the four different kinds of Food prepared in the kitchen will be served.

Near the eating-rooms, other rooms will be neatly fitted up, and kept constantly clean, and well warmed; and well lighted in the evening; in which the Poor who frequent the Establishment will be permitted to remain during the day, and till a certain hour at night. — They will be allowed and even ENCOURAGED to bring their work with them to these rooms; and by degrees they will be furnished with utensils, and raw materials for working for their own emolument, by the Establishment. Praises and rewards will be bestowed on those who most distinguish themselves by their industry, and by their peaceable and orderly behaviour.

In the fitting up of the kitchen, care will be taken to introduce every useful invention and improvement, by which fuel may be saved, and the various processes of cookery facilitated, and rendered less expensive; and the whole mechanical arrangement will be made as complete and perfect as possible, in order that it may serve as a model for imitation; and care will be likewise be taken in fitting up the dining-halls, and other rooms belonging to the Establishment, to introduce the most approved fire-places, — stoves, — flews, and other mechanical contrivances for heating rooms and passages; — as also in lighting up the house to make use of a variety of the best, most economical, and most beautiful lamps; and in short, to collect together such an assemblage of useful and elegant inventions, in every part of the Establishment, as to render it not only an object of public curiosity, but also of the most essential and extensive utility.

And although it will not be possible to make the Establishment sufficiently extensive to accommodate all the Poor of so large a city, yet it may easily be made large enough to afford a comfortable asylum to a great number of distressed objects; and the interesting and affecting scene it will afford to spectators, can hardly fail to attract the curiosity of the Public; and there is great reason to hope that the success of the experiment, and the evident tendency of the measures adopted to promote the comfort, happiness, and prosperity of society, will induce many to exert themselves in forming similar Establishments in other places. — It is even probable that the success which will attend this first essay, (for successful it must, and will be, as care will be taken to limit its extent to the means furnished for carrying it into execution,) will encourage others, who do not put down their names upon the lists of the subscribers at first, to follow with subscription for the purpose of augmenting the Establishment, and rendering it more extensively useful.

Should this be the case, it is possible that in a short time subordinate public kitchens, with rooms adjoining them for the accommodation of the industrious Poor, may be established in all the parishes;—and when this is done, only one short step more will be necessary in order to complete in the management of the Poor. Poor rates may then be entirely abolished, and VOLUNTARY SUBSCRIPTIONS, which certainly need never amount to one half what the Poor rates now are, may be substituted in the room of them, and one general establishment may be formed for the relief and support of the Poor in this capital.

It will however be remembered that it is by no means the intention of the Author of these Proposals that those who contribute to the object immediately in view, the forming A MODEL for an Establishment for feeding and giving employment to the Poor, should be troubled with any future solicitations on that score; very far from it, measures will be so taken, by limiting the extent of the undertaking to the amount of the sums subscribed, and by arranging matters so that the Establishment. once formed, shall be able to support itself, that no farther assistance from the subscribers will be necessary.—If any of them should, of their accord, follow up their subscription by other donations, these additional sums will be thankfully received, and faithfully applied, to the general or particular purposes for which they may be designed; but the subscribers may depend upon never being troubled with any future SOLICITATIONS on any pretence whatever, on account of the present undertaking.

A secondary object in forming this Establishment, and which will be attended to as soon as the measures for feeding the Poor, and giving them employment, are carried into execution, is the forming of a grand repository of all kinds of USEFUL MECHANICAL INVENTIONS, and particularly of such as relate to the furnishing of houses, and are calculated to promote domestic comfort and economy.

Such a repository will not only be highly interesting, considered as an object of public curiosity, but it will be really useful, and will doubtless contribute very powerfully to the introduction of many essential improvements.

To render this part of the Establishment still more complete, rooms will be set apart for receiving, and exposing to public view, all such new and useful inventions as shall, from time to time, be made, in this, or in any other country, and sent to the institution; and a written account, containing the name of the inventor,—the place where the article may be bought,—and the price of it, will be attached to each article, for the information of those who may be desirous of knowing any of these particulars.

If the amount of the subscriptions should be sufficient to defray the additional expence which such an arrangement would require, models will be prepared, upon a reduced scale, for showing the improvements which may be made in the construction of the coppers, or boilers, used by brewers, and distillers, as also of their fire-places; with a view both to the economy of the fuel, and to convenience.

Complete kitchens will likewise be constructed, of the full size, with all their utensils, as models for private families.— And that these kitchens may not be useless, eating rooms may be fitted up adjoining to them, and cooks engaged to furnish to gentlemen, subscribers, or others, to whom subscribers may delegate that right, good dinners, at the prime cost of the victuals, and the expense of cooking, which together certainly would not exceed ONE SHILLING A HEAD.

The public kitchen from whence the Poor will be fed will be so constructed as to serve as a model for hospitals, and for other great Establishments of similar nature.

The expense of feeding the Poor will be provided for by selling the portions of Food delivered from the public kitchen at such a price, that those expenses shall be just covered, and no more:— so that the Establishment, when once completed, will be made to support itself.

Tickets for Food (which may be considered as drafts upon the public kitchen, payable at sight) will be furnished to all persons who apply for them, in as far as it shall be possible to supply the demands; but care will be taken to provide, first, for the Poor who frequent regularly the working rooms belonging to the Establishment; and secondly, to pay attention to the recommendations of subscribers, by furnishing Food immediately, or with the least possible delay, to those who come with subscribers' tickets.

As soon as the Establishment shall be completed, every subscriber will be furnished gratis with tickets for Food, to the amount of ten per cent. of his subscription; the value of the tickets being reckoned at what the portions of Food really cost, which will be delivered to those who produce the tickets at the public kitchen. — At the end of six months, tickets to the amount of ten per cent. more, and so on, at the end of every six succeeding months, tickets to the amount of ten per cent. of the sum subscribed will be delivered to each subscriber till he shall actually have received in tickets for Food, or drafts upon the public kitchen, to the full amount of ONE HALF of his original subscription. — And as the price at which this Food will be charged, will be at the most moderate computation, at least FIFTY PER CENT. cheaper than it would cost any where else, the subscribers will in fact receive in these tickets the full value of the sums they will have subscribed; so that in the end, the whole advance will be repaid, and a most interesting, and most useful public institution will be completely established WITHOUT ANY EXPENSE TO ANYBODY — And the Author of these Proposals will think himself most amply repaid for any trouble he may have in the execution of this scheme, by the heartfelt satisfaction he will enjoy in the reflection of having been instrumental in doing essential service to mankind.

It is hardly necessary to add, that although the subscribers will receive in return for their subscriptions the full value of them, in tickets, or orders upon the public kitchen, for Food, yet the property of the Whole Establishment, with all its appurtenances, will nevertheless remain vested solely and entirely in the subscribers, and their lawful heirs; and that they will have power to dispose of it in any way they may think proper, as also to give orders and directions for its future management. (Signed) "A. B." London, 1st January 1796.

These Proposals, which should be printed, and distributed gratis, in great abundance, should be accompanied with subscription-lists which should be printed on fine writing-paper; and to save trouble to the subscribers, might be of a peculiar form. — Upon the top of a half-sheet of folio writing-paper might be printed, the following Head of Title, and the remainder of that side of the half-sheet, below this Head, might be formed into different columns, thus:

SUBSCRIPTIONS,

For carrying into execution the scheme for forming an Establishment for feeding the Poor from a Public KITCHEN, and giving them useful employment, etc. proposed by A. B. and particularly described in the printed paper, dated London, 1st January 1796, which accompanies this Subscription List.

N.B. No part of the money subscribed will be called for, unless it be found that the amount of the subscriptions will be quite sufficient to carry the scheme proposed into complete execution without troubling the subscribers a second time for further assistance.

— —--

Subscribers Names. I Place of Abode. I Sums subscribed.

— —--

      I I pound. s. d.
      I I
      I I
      I I
      I I
      I I
      I I

that this list is authentic, and that the persons mentioned in it have agreed to subscribe the sums placed against their names, is attested by [ ]. The person who is so good as to take charge of this list, is requested to authenticate it by signing the above certificate, and then to seal it up and send it according to the printed address on the back of it.

The address upon the back of the subscription lists, (which may be that of the author of the Proposal, or of any other person he may appoint to receive these lists,) should be printed in such a manner that, when the list is folded up in the form of a letter, the address may be in its proper place. This will save trouble to those who take charge of these lists; and too much pains cannot be taken to give as little trouble as possible to persons who are solicited to contribute IN MONEY towards carrying into execution schemes of public utility.

As a public Establishment like that here proposed would be highly interesting, even were it to be considered in no other light than merely as an object of curiosity, there is no doubt but it would be much frequented; and it is possible that this concourse of people might be so great as to render it necessary to make some regulations in regard to admittance: but, whatever measures might be adopted with respect to others, SUBSCRIBERS ought certainly to have free admittance at all times to every part of the Establishment,—They should even have a right individually to examine all the details of its administration, and to require from those employed as overseers, or managers, any information or explanation they might want.— They ought likewise to be at liberty to take drawings, or to have them taken by others, (at their expense,) for themselves or for their friends, of the kitchen, stoves, grates, furniture, etc. and in general of every part of the machinery belonging to the Establishment.

In forming the Establishment, and providing the various machinery, care should be taken to employ the most ingenious and most respectable tradesmen; and if the name of the maker, and the place of his abode were to be engraved or written on each article, this, no doubt, would tend to excite emulation among the artizans, and induce them to furnish goods of the best quality, and at as low a price as possible.—It is even possible, that in a great and opulent city like London, and where public spirit and zeal for improvement pervade all ranks of society, many respectable tradesmen in easy circumstances might be found, who would have real pleasure in furnishing gratis such of the articles wanted as are in their line of business: and the advantages which might, with proper management, be derived from this source, would most probably be very considerable.

With regard to the management of the Poor who might be collected together for the purpose of being fed and furnished with employment, in a Public Establishment like that here recommended, I cannot do better than refer my reader to the account already published (in my First Essay) of the manner in which the Poor at Munich were treated in the house of industry established in that city, and the means that were used to render them comfortable, HAPPY, and industrious.

As soon as the scheme here recommended is carried into execution, and measures are effectually taken for feeding the Poor at a cheap rate, and giving them useful employment, no farther difficulties will then remain, at least none certainly that are insurmountable, to prevent the introduction of a general plan for providing for all the Poor, founded upon the principles explained and recommended in the preceding Chapters of this Essay.

CHAPTER. V.

Of the Means which may be used by Individuals in affluent
  Circumstances for the Relief of the Poor in their Neighbourhood.

As nothing tends more powerfully to encourage idleness and immorality among the Poor, and consequently to perpetuate all the evils to society which arise from the prevalence of poverty and mendicity, than injudicious distributions of alms; individuals must be very cautious in bestowing their private charities, and in forming schemes for giving assistance to the distressed; otherwise they will most certainly do more harm than good. — The evil tendency of giving alms indiscriminately to beggars is universally acknowledged; but it is not, I believe, so generally known how much harm is done by what are called the PRIVATE CHARITIES of individuals. — Far be it from me to wish to discourage private charities; I am only anxious that they should be better applied.

Without taking up time in analyzing the different motives by which persons of various character are induced to give alms to the Poor, or of showing the consequences of their injudicious or careless donations; which would be an unprofitable as well as a disagreeable investigation; I shall briefly point out what appear to me to be the most effectual means which individuals in affluent circumstances can employ for the assistance of the Poor in their neighbourhood.

The most certain and efficacious relief that can be given to the Poor is that which would be afforded them by forming a general Establishment for giving them useful employment, and furnishing them with the necessaries of life at a cheap rate; in short, forming a Public Establishment similar in all respects to that already recommended, and making it as extensive as circumstances will permit.

An experiment might first be made in a single village, or in a single parish; a small house, or two or three rooms only, might be fitted up for the reception of the Poor, and particularly of the children of the Poor; and to prevent the bad impressions which are sometimes made by names which have been become odious, instead of calling it a Work-house, it might be called "A School of Industry," or, perhaps, Asylum would be a better name for it. — One of these rooms should be fitted up as a kitchen for cooking for the Poor; and a middle-aged women of respectable character, and above all of a gentle and humane disposition, should be placed at the head of this Establishment, and lodged in the house. — As she should serve at the same time as chief cook, and as steward of the institution, it would be necessary that she should be able to write and keep accounts; and in cases where the business of superintending the various details of the Establishment would be too extensive to be performed by one person, one or more assistants may be given her.

In large Establishments it might, perhaps, be best to place a married couple, rather advanced in life, and without children, at the head of the institution; but, whoever are employed in that situation, care should be taken that they should be persons of irreproachable character, and such as the Poor can have no reason to suspect of partiality.

As nothing would tend more effectually to ruin an Establishment of this kind, and prevent the good intended to be produced by it, than the personal dislikes of the Poor to those put over them, and more especially such dislikes as are founded on their suspicions of their partiality, the greatest caution in the choice of these persons will always be necessary: and in general it will be best not to take them from among the Poor, or at least not from among those of the neighbourhood, nor such as have relations, acquaintances, or other connexions among them.

Another point to be attended to in the choice of a person to be placed at the head of such an Establishment, (and it is a point of more importance than can well be imagined by those who have not considered the matter with some attention)— is the looks or EXTERNAL APPEARANCE of the person destined for this employment.

All those who have studied human nature, or have taken notice of what passes in themselves when they approach for the first time a person who has any thing very strongly marked in his countenance, will feel how very important it is that a person placed at the head of an asylum for the reception of the Poor and the unfortunate should have an open, pleasing countenance, such as inspires confidence and conciliates affection and esteem.

Those who are in distress, are apt to be fearful and apprehensive, and nothing would be so likely to intimidate and discourage them as the forbidding aspect of a stern and austere countenance in the person they were taught to look up to for assistance and protection.

The external appearance of those who are destined to command others is always a matter of real importance, but it is peculiarly so when those to be commanded and directed are objects of pity and commiseration.

Where there are several gentlemen who live in the neighbourhood of the same town or village where an Establishment, or Asylum, (as I would wish it might be called,) for the Poor is to be formed, they should all unite to form ONE ESTABLISHMENT, instead of each forming a separate one; and it will likewise be very useful in all cases to invite all ranks of people resident within the limits of the district in which an Establishment is formed, except those who are actually in need of assistance themselves, to contribute to carry into execution such a public undertaking; for though the sums the more indigent and necessitous of the inhabitants may be able to spare may be trifling, yet their being invited to take part in so laudable an undertaking will be flattering to them, and the sums they contribute, however small they may be, will give them a sort of property in the Establishment, and will effectually engage their good wishes at least, (which are of more importance in such cases than is generally imagined,) for its success.

How far the relief which the Poor would receive from the execution of a scheme like that here proposed ought to preclude them from a participation of other public charities, (in the distribution of the sums levied upon the inhabitants in Poor's taxes, for instance, where such exist,) must be determined in each particular case according to the existing circumstances. It will, however, always be indispensably necessary where the same poor person receives charitable assistance from two or more separate institutions, or from two or more private individuals, at the same time, for each to know exactly the amount of what the others give, otherwise too much or too little may be given, and both tend to discourage INDUSTRY, the only source of effectual relief to the distresses and the misery of the Poor. — And hence may again be seen the great importance of what I have so often insisted on, the rendering of measures for the relief of the Poor as general as possible.

To illustrate in the clearest manner, and in as few words as possible, the plan I would recommend for forming an Establishment for the Poor on a small scale — such as any individual even of moderate property, might easily execute; I will suppose that a gentleman, resident in the country upon his own estate, has come to a resolution to form such an Establishment in a village near his house, and will endeavour briefly to point out the various steps he would probably find it necessary to take in the execution of this benevolent and most useful undertaking.

He would begin by calling together at his house the clergyman of the parish, overseers of the Poor, and other parish officers, to acquaint them with his intentions, and ask their assistance and friendly co-operation in the prosecution of the plan; the details of which he would communicate to them as far as he should think it prudent and necessary at the first outset to entrust them indiscriminately with that information. — The characters of the persons, and the private interest they might have to promote or oppose the measures intended to be pursued, would decide upon the degree of confidence which ought to be given them.

At this meeting, measures should be taken for forming the most complete and most accurate lists of all the Poor resident within the limits proposed to be given to the Establishment, with a detailed account of every circumstance, relative to their situation, and their wants. — Much time and trouble will be saved in making out these lists, by using printed forms or blanks similar to those made use of at Munich; and these printed forms will likewise contribute very essentially to preserve order and to facilitate business, in the management of a private as well as of a public charity; — as also to prevent the effects of misrepresentation and partiality on the part of those who must necessarily be employed in these details.

Convenient forms or models for these blanks will be given in the Appendix to this volume.

At this meeting, measures may be taken for numbering all the houses in the village or district, and for setting on foot private subscriptions among the inhabitants for carrying the proposed scheme into execution.

Those who are invited to subscribed should be made acquainted, by a printed address accompanying the subscription lists, with the nature, extent, and tendency of the measures adopted; and should be assured that, as soon as the undertaking shall be completed, the Poor will not only be relieved, and their situation made more comfortable, but mendicity will be effectually prevented, and at the same time the Poor's rates, or the expense to the public for the support of the Poor, very considerably lessened.

These assurances, which will be the strongest inducements that can be used to prevail on the inhabitants of all descriptions to enter warmly into the scheme, and assist with alacrity in carrying it into execution, should be expressed in the strongest terms; and all persons of every denomination, young and old, and of both sexes, (paupers only excepted,) should be invited to put down their names in the subscription lists, and this even, HOWEVER SMALL THE SUMS MAY BE WHICH THEY ARE ABLE TO CONTRIBUTE. — Although the sums which day-labourers, servants, and other in indigent circumstances may be able to contribute, may be very trifling, yet there is one important reason why they ought always to be engaged to put down their names upon the lists as subscribers, and that is the goods effects which their taking an active part in the undertaking will probably produce ON THEMSELVES.—Nothing tends more to mend the heart, and awaken in the mind a regard for character, than acts of charity and benevolence; and any person who has once felt that honest pride and satisfaction which result from a consciousness of having been instrumental in doing good by relieving the wants of the Poor, will be rendered doubly careful to avoid the humiliation of becoming himself an object of public charity.

It was a consideration of these salutary effects, which may always be expected to be produced upon the minds of those who take an active and VOLUNTARY part in the measures adopted for the relief of the Poor, that made me prefer voluntary subscriptions, to taxes, in raising the sums necessary for the support of the Poor, and all the experience I have had in these matters has tended to confirm me in the opinion I have always had of their superior utility,—Not only day-labourers and domestic servants, but their young children, and all the children of the nobility and other inhabitants of Munich, and even the non-commissioned officers and private soldiers of the regiments in garrison in that city, were invited to contribute to the support of the institution for the Poor; and there are very few indeed of any age or condition (paupers only excepted) whose names are not to be found on the lists of subscribers.

The subscriptions at Munich are by families, as has elsewhere been observed; and this method I would recommend in the case under consideration, and in all others.—The head of the family takes the trouble to collect all the sums subscribed upon his family list, and to pay them into the hands of those who (on the part of the institution) are sent round on the first Sunday morning of every month to receive them; but the names of all the individuals who compose the family are entered on the list at full length, with the sum each contributes.

Two lists of the same tenor must be made out for each family; one of which must be kept by the head of the family for his information and direction, and the other sent in to those who have the general direction of the Establishment.

These subscription-lists should be printed; and they should be carried round and left with the heads of families, either by the person himself who undertakes to form the Establishment, (which will always be best,) or at least by his steward, or some other person of some consequence belonging to his household. —Forms or models for these lists may be seen in the Appendix.

When these lists are returned, the person who has undertaken to form the Establishment will see what pecuniary assistance he is to expect; and he will either arrange his plan, or determine the sum he may think proper to contribute himself, according to that amount.— He will likewise consider how far it will be possible and ADVISABLE to connect his scheme with any Establishment for the relief of the Poor already existing; or to act in concert with those in whose hands the management of the Poor is vested by the laws.—These circumstances are all important; and the manner of proceeding in carrying the proposed scheme into execution must, in a great measure, be determined by them. Nothing, however, can prevent the undertaking from being finally successful, provided the means used for making it so are adopted with caution, and pursued with perseverance.

However adverse those may be to the scheme who, were they well disposed, could most effectually contribute to its success—yet no opposition which can be given to it by INTERESTED PERSONS,— such as find means to derive profit to themselves in the administration of the affairs of the Poor;—no opposition, I say, from such persons, (and none surely but these can ever be desirous of opposing it,) can prevent the success of a measure so evidently calculated to increase the comforts and enjoyments of the Poor, and to promote the general good of society.

If the overseers of the Poor, and other parish officers, and a large majority of the principal inhabitants, could be made to enter warmly into the scheme, it might, and certainly would, in many cases, be possible, even without any new laws or acts of parliament being necessary to authorize the undertaking, to substitute the arrangements proposed in the place of the old method of providing for the Poor;—abolishing entirely, or in so far as it should be found necessary,—the old system, and carrying the scheme proposed into execution as a GENERAL MEASURE.

In all cases where this can be effected, it ought certainly to be preferred to any private or less general institution; and individuals, who, by their exertions, are instrumental in bringing about so useful a change, will render a very essential service to society:—But even in cases where it would not be possible to carry the scheme proposed into execution in its fullest extent, much good may be done by individuals in affluent circumstances to the Poor, by forming PRIVATE ESTABLISHMENTS for feeding them and giving them employment.

Much relief may likewise be afforded them by laying in a large stock of fuel, purchased when it is cheap, and retailing it out to them in small quantities, in times of scarcity, at the prime cost.

It is hardly to be believed how much the Poor of Munich have been benefited by the Establishment of the Wood-magazine, from whence they are furnished in winter, during the severe frosts, with fire-wood at the price it costs when purchased in summer, in large quantities, and at the cheapest rate. And this arrangement may easily be adopted in all countries, and by private individuals as well as by communities. Stores may likewise be laid in of potatoes, peas, beans, and other articles of food, to be distributed to the Poor in like manner, in small quantities, and at low prices; which will be a great relief to them in times of scarcity. It will hardly be necessary for me to observe, that in administering this kind of relief to the Poor it will often be necessary to take precautions to prevent abuses.

Another way in which private individuals may greatly assist the Poor, is, by showing them how they may make themselves more comfortable in their dwellings. Nothing is more perfectly miserable and comfortless than the domestic arrangement of poor families in general; they seem to have no idea whatever of order or economy in any thing; and every thing about them is dreary, sad, and neglected, in the extreme. A little attention to order and arrangement would contribute greatly to their comfort and conveniences, and also to economy. They ought in particular to be shown how to keep their habitations warm in winter, and to economise fuel, as well in heating their rooms, as in cooking, washing, etc.

It is not to be believed what the waste of fuel really is, in the various processes in which it is employed in the economy of human life; and in no case is this waste greater than in the domestic management of the Poor. Their fire-places are in general constructed upon the most wretched principles; and the fuel they consume in them, instead of heating their rooms, not unfrequently renders them really colder, and more uncomfortable, by causing strong currents of cold air to flow in from all the doors and windows to the chimney. This imperfection of their fire-places may be effectually remedied;—these currents of cold air prevented,—above half their fuel saved,—and their dwellings made infinitely more comfortable, merely by diminishing their fire-places, and the throats of their chimnies just above the mantle-piece; which may be done as a very every trifling expence, with a few bricks, or stones, and a little mortar, by the most ordinary bricklayer. And with regard to the expence of fuel for cooking, so simple a contrivance as an earthen pot, broad at top, for receiving a stew-pan, or kettle, and narrow at bottom, with holes through its sides near the bottom, for letting in air under a small circular iron grate, and other small holes near the top for letting out the smoke, may be introduced with great advantage. By making use of this little portable furnace, (which is equally well adapted to burn wood, or coals.)—one eighth part of the fuel will be sufficient for cooking, which would be required were the kettle to be boiled over an open fire.—To strengthen this portable furnace, it may be hooped with iron hoops, or bound round with strong iron wire:—but I forget that I am anticipating the subject of a future Essay.

Much good may also be done to the Poor by teaching them how to prepare various kinds of cheap and wholesome food, and to render them savoury and palatable.—The art of cookery, notwithstanding its infinite importance to mankind, has hitherto been little studied; and among the more indigent classes of society, where it is most necessary to cultivate it, it seems to have been most neglected.—No present that could be made to a poor family could be of more essential service to them than a thin, light stew-pan, with its cover, made of wrought, or cast iron, and fitted to a portable furnace, or close fire-place, constructed to save fuel; with two or three approved receipts for making nourishing and savoury soups and broths at a small expence.

Such a present might alone be sufficient to relieve a poor family from all their distresses, and make them permanently comfortable; for the expences of a poor family for food might, I am persuaded, in most cases be diminished ONE HALF by a proper attention to cookery, and to the economy of fuel; and the change in the circumstances of such a family, which would be produced by reducing their expenses for food to one half what it was before, is easier to be conceived than described.

It would hardly fail to re-animate the courage of the most desponding;—to cheer their drooping spirits, and stimulate them to fresh exertions in the pursuits of useful industry.

As the only effectual means of putting an end to the sufferings of the Poor is the introduction of a spirit of industry among them, individuals should never lose sight of that great and important object, in all the measures they may adopt to relieve them.—But in endeavouring to make the Poor industrious, the utmost caution will be necessary to prevent their being disgusted.—Their minds are commonly in a state of great irritation, the natural consequences of their sufferings, and of their hopeless situation; and their suspicions of every body about them, and particularly of those who are set over them, are so deeply rooted that it is sometimes extremely difficult to sooth and calm the agitation of their minds, and gain their confidence. —This can be soonest and most effectually done by kind and gentle usage; and I am clearly of opinion that no other means should ever be used, except it be with such hardened and incorrigible wretches as are not to be reclaimed by any means; but of these, I believe, there are very few indeed.—I have never yet found one, in all the course of my experience in taking care of the Poor.

We have sometimes been obliged to threaten the most idle and profligate with the house of correction; but these threats, added to the fear of being banished from the House of Industry, which has always been held up and considered as the greatest punishment, have commonly been sufficient for keeping the unruly in order.

If the force of example is irresistible in debauching men's minds, and leading them into profligate and vicious courses, it is not less so in reclaiming them, and rendering them orderly, docile, and industrious; and hence the infinite importance of collecting the Poor together in Public Establishments, where every thing about them is animated by unaffected cheerfulness, and by that pleasing gaiety, and air of content and satisfaction, which always enliven the busy scenes of useful industry.

I do not believe it would be possible for any person to be idle in the House of Industry at Munich. I never saw any one idle; often as I have passed through the working-rooms; nor did I ever see any one to whom the employments of industry seemed to be painful or irksome.

Those who are collected together in the public rooms destined for the reception and accommodation of the Poor in the day-time, will not need to be forced, nor even urged to work;—if there are in the room several persons who are busily employed in the cheerful occupations of industry, and if implements and materials for working are at hand, all the others present will not fail to be soon drawn into the vortex, and joining with alacrity in the active scene, their dislike to labour will be forgotten, and they will become by habit truly and permanently industrious.

Such is the irresistible power of example! — Those who know how to manage this mighty engine and have opportunities of employing it with effect, may produce the most miraculous changes, in the manners, disposition, and character, even of whole nations.

In furnishing raw materials to the Poor to work, it will be necessary to use many precautions to prevent frauds and abuses, not only on the part of the Poor, who are often but too much disposed to cheat and deceive whenever they find opportunities, but also on the part of those employed in the details of this business: — but the fullest information having already been given in my First Essay, of all the various precautions it had been found necessary to take for the purposes in question in the House of Industry at Munich, it is not necessary for me to enlarge upon the subject in this place, or to repeat what has already been said upon it elsewhere.

With regard to the manner in which good and wholesome food for feeding the Poor may be prepared in a public kitchen, at a cheap rate, I must refer my reader to my Essay on Food; where he will find all the information on that subject which he can require. —In my Essay on Clothing, he will see how good and comfortable clothing may be furnished to the Poor at a very moderate expence; and in that on the Management of Heat, he will find particular directions for the Poor for saving fuel.

I cannot finish this Essay, without taking notice of a difficulty which frequently occur in giving employment to the Poor, that of disposing to advantage of the produce of their labour: — This is in all cases a very important object; and too much attention cannot be paid to it. — A spirit of industry cannot be kept up by making it advantageous to individuals to be industrious; but where the wages which the labourer has a right to expect are refused, it will not be possible to prevent his being discouraged and disgusted. — He may perhaps be forced for a certain time to work for small wages, to prevent starving, if he has not the resource of throwing himself upon the parish, which he most probably would prefer doing, should it be in his option; but he will infallibly conceive such a thorough dislike to labour, that he will become idle and vicious, and a permanent and heavy burden on the public.

If "a labourer is worthy of his hire," he is peculiarly so, where that labourer is a poor person, who, with all his exertions, can barely procure the first necessaries of life; and whose hard lot renders him an object of pity and compassion.

The deplorable situation of a poor family, struggling with poverty and want, — deprived of all the comforts and conveniences of life — deprived even of hope; and suffering at the same time from hunger, disease, and mortifying and cruel disappointment, is seldom considered with that attention which it deserves, by those who have never felt these distresses, and who are not in danger of being exposed to them. My reader must pardon me, if I frequently recall his attention to these scenes of misery and wretchedness. He must be made acquainted with the real situation of the Poor — with the extent and magnitude of their misfortunes and sufferings, before it can be expected that he should enter warmly into measures calculated for their relief. In forming Establishments, public or private, for giving employment to the Poor, it will always be indispensably necessary to make such arrangements as will secure to them a fair price for all the labour they perform. They should not be OVER-PAID, for that would be opening a door for abuse; — but they ought to be generously paid for their work; and, above all, they ought never to be allowed to be idle for the want of employment. The kind of employment it may be proper to give them will depend much on local circumstances. It will depend on the habits of the Poor; — the kinds of work they are acquainted with; — and the facility with which the articles they can manufacture may be disposed of at a good price.

In very extensive Establishments, there will be little difficulty in finding useful employment for the Poor; for where the number of persons to be employed is very great, a great variety of different manufactures may be carried on with advantage, and all the articles manufactured, or prepared to be employed in the manufactures, may be turned to a good account.

In a small Establishment, circumscribed and confined to the limits of a single village or parish, it might perhaps be difficult to find a good market for the yarn spun by the Poor; but in a general Establishment, extending over a whole country, or large city, as the quantity of yarn spun by all the Poor within the extensive limits of the institution will be sufficient to employ constantly a number of weavers of different kinds of cloth and stuff, the market for all the various kinds of yarn the Poor may spin will always be certain. The same reasoning will hold with regard to various other articles used in great manufactories, upon which the Poor might be very usefully employed; and hence the great advantage of making Establishments for giving employment to the Poor as extensive as possible. It is what I have often insisted on, and what I cannot too strongly recommend to all those who engage in forming such Establishments.

Although I certainly should not propose to BRING TOGETHER, under one roof, all the Poor of a whole kingdom, as, by the inscription over the entrance into a vast hospital began, but not finished, at Naples, it would appear was once the intention of the government in that country; yet I am clearly of opinion that an institution for GIVING EMPLOYMENT TO THE POOR can hardly be too extensive.

But to return to the subject to which this Chapter was more particularly appropriated, the relief that may be afforded by private individuals to the Poor in their neighbourhood; in case it should not be possible to get over all the difficulties that may be in the way to prevent the forming of a general Establishment for the benefit of the Poor, individuals must content themselves with making such private arrangements for that purpose as they may be able, WITH SUCH ASSISTANCE AS THEY CAN COMMAND, to carry into execution.

The most simple, and least expensive measure that can be adopted for the assistance of the Poor will be that of furnishing them with raw materials for working. Flax, hemp, or wool, for instance, for spinning; and paying them in money, at the market price, for the yarn spun. This yarn may afterwards be sent to weavers to be manufactured into cloth, or may be sent to some good market and sold. The details of these mercantile transactions will be neither complicated nor troublesome, and might easily be managed by a steward of house-keeper; particularly if the printed tickets, and tables, I have so often had occasion to recommend, are used.

The flax, hemp, or wool, as soon as it is purchased, should be weighed out into bundles of one or two pounds each, and lodged in a store-room; and when one of these bundles is delivered out to a poor person to be spun, it should be accompanied with a printed spin-ticket, and entered in a table to be kept for that purpose; and when it is returned spun, an abstract of the spin-ticket itself, should be bound up with the bundle of yarn, in order that any frauds committed by the spinner, in reeling, or in any other way, which may be discovered upon winding off the yarn, may be brought home to the person who committed them. When it is known that such effectual precautions to detect frauds are used, no farther attempts will be made to defraud; and a most important point indeed will be gained, and one which will most powerfully tend to mend the morals of the Poor, and restore peace to their minds. When, by rendering it evidently impossible for them to escape detection, they are brought to give up all thoughts of cheating and deceiving; they will then be capable of application, and of enjoying real happiness, and, with open and placid countenances, will look every one full in the face who accosts them: but as long as they are under the influence of temptation —as long as their minds are degraded by conscious guilt, and continually agitated by schemes of prosecuting their fraudulent practices, they are as incapable of enjoying peace or contentment, as they are of being useful members of society.

Hence the extreme cruelty of an ill-judged appearance of confidence, or careless neglect of precautions, in regard to those employed in places of trust, who may be exposed to temptations to defraud.

That prayer, which cannot be enough admired, or too often repeated, "LEAD US NOT INTO TEMPTATION," was certainly dictated by infinite wisdom and goodness; and it should ever be borne in mind by those who are placed in stations of power and authority, and whose measures must necessarily have much influence on the happiness or misery of great numbers of people.

Honest men may be found in all countries; but I am sorry to say, that the result of all my experience and observation has tended invariably to prove, (what has often been remarked,) that it is extremely difficult to KEEP THOSE HONEST who are exposed to continual and great temptations.

There is, however, one most effectual way, not only of keeping those honest who are so already, but also of making those honest who are not so; and that is, by taking such precautions as will render it EVIDENTLY impossible for those who commit frauds to escape detection and punishment: and these precautions are never impossible, and seldom difficult; and with a little address, they may always be so taken as to be in nowise offensive to those who are the objects of them.

It is evident that the maxims and measures here recommended are not applicable merely to the Poor, but also, and more especially, to those who may be employed in the details of relieving them.

But to return once more to the subject more immediately under consideration. — If individuals should extend their liberality so far as to establish public kitchens for feeding the Poor, (which is a measure I cannot too often, or too forcibly recommend,) it would be a great pity not to go one easy step further, and fit up a few rooms adjoining to the kitchen, where the Poor may be permitted to assemble to work for their own emoluments, and where schools for instructing the children of the Poor in working, and in reading and writing, may be established. Neither the fitting up, or warming and lighting of these rooms, will be attended with any considerable expense; while the advantages which will be derived from such an Establishment for encouraging industry, and contributing to the comfort of the Poor, will be most important; and from their peculiar nature, and tendency, will be most highly interesting to every benevolent mind.

END OF ESSAY TWO.

Footnotes for Essay II.

[1] This English Reader is desired to bear in mind, that the Author of this Essay, though an Englishman, is resident in Germany; and that his connections with that country render it necessary for him to

pay particular attention to its circumstances, in treating a subject which he is desirous of rendering generally useful. These is still another reason, which renders it necessary for him to have continually in view, in the Treatise, the situation of the Poor upon the Continent, and that it is an engagement which he has laid himself under to write upon that subject.

[2] The only step which, in my opinion, it would be either, necessary, or prudent, for the legislature to take in any country where an Establishment for the Poor is to be formed, is to RECOMMEND to the Public a good plan for such an Establishment, and repeal, or alter all such of the existing laws as might render the introduction of it difficult or impossible.

[3] This is an object of the utmost importance, and the success of the undertaking will depend in a great measure on the attention that is paid to it.

[4] This measure has been followed by the most salutary effects at Munich. The commissaries of districts flattered by this distinction have exerted themselves with uncommon zeal and assiduity in the discharge of the important duties of their office. And very important indeed is the office of a commissary of a district in the Establishment for the Poor at Munich.

[5] It will be best, if it be possible, to mention and describe the place, in the Proposals.

CONTENTS of ESSAY III.

of FOOD and particularly of FEEDING the POOR

Introduction.

CHAPTER. I. Great importance of the subject under consideration. Probability that water acts a much more important part in nutrition than has hitherto been generally imagined. Surprisingly small quantity of solid food necessary, when properly prepared, for all the purposes of nutrition. Great importance of the art of cookery. Barley remarkably nutritive when properly prepared. The importance of culinary processes for preparing food shown

ing cheap soups. Receipt for the cheapest soup that can be made. Of SAMP Method of preparing it Is an excellent Substitute for Bread. Of brown Soup. Of RYE BREAD.

ESSAY III.

INTRODUCTION.

It is a common saying, that necessity is the mother of invention; and nothing is more strictly or more generally true. It may even be shown, that most of the successive improvements in the affairs of men in a state of civil society, of which we have any authentic records, have been made under the pressure of necessity; and it is no small consolation, in times of general alarm, to reflect upon the probability that, upon such occasions, useful discoveries will result from the united exertions of those who, either from motives of fear, or sentiments of benevolence, labour to avert the impending evil.

The alarm in this country at the present period[1], on account of the high price of corn, and the danger of a scarcity, has turned the attention of the Public to a very important subject, THE INVESTIGATION OF THE SCIENCE OF NUTRITION;—a subject so curious in itself, and so highly interesting to mankind, that it seems truly astonishing it should have been so long neglected:— but in the manner in which it is now taken up, both by the House of Commons, and the Board of Agriculture, there is great reason to hope that it will receive a thorough scientific examination; and if this should be the case, I will venture to predict, that the important discoveries, and improvements, which must result from these enquiries, will render the alarms which gave rise to them for ever famous in the annals of civil society.

CHAPTER. I.

Great importance of the subject under consideration.
Probability that water acts a much more important part in
  nutrition than has hitherto been generally imagined.
Surprisingly small quantity of solid food necessary,
  when properly prepared, for all the purposes of nutrition.
Great importance of the art of cookery.
Barley remarkably nutritive when properly prepared.
The importance of culinary processes for preparing food shown
  from the known utility of a practice common in some parts of
  Germany of cooking for cattle.
Difficulty of introducing a charge of cookery into common use.
Means that may be employed for that purpose.

There is, perhaps, no operation of Nature, which falls under the cognizance of our senses, more surprising, or more curious, than the nourishment and growth of plants, and animals; and there is certainly no subject of investigation more interesting to mankind. — As providing subsistence is, and ever must be, an object of the first concern in all countries, any discovery or improvement by which the procuring of good and wholesome food can be facilitated, must contribute very powerfully to increase the comforts, and promote the happiness of society.

That our knowledge in regard to the science of nutrition is still very imperfect, is certain; but, I think there is reason to believe, that we are upon the eve of some very important discoveries relative to that mysterious operation.

Since it has been known that Water is not a simple element, but a COMPOUND, and capable of being decomposed, much light has been thrown upon many operations of nature which formerly were wrapped up in obscurity. In vegetation, for instance, it has been rendered extremely probable, that water acts a much more important part than was formerly assigned to it by philosophers. — That it serves not merely as the VEHICLE of nourishment, but constitutes at least one part, and probably an essential part, of the

FOOD of plants.—That it is decomposed by them, and contributes MATERIALLY to their growth;—and that manures serve rather to prepare the water for decomposition, than to form of themselves—substantially, and directly—the nourishment of the vegetables.

Now, a very clear analogy may be traced, between the vegetation and growth of plants, and the digestion and nourishment of animals; and as water is indispensably necessary in both processes, and as in one of them, (vegetation,) it appears evidently to serve as FOOD;—why should we not suppose it may serve as food in the other?—There is, in my opinion, abundant reason to suspect that this is really the case; and I shall now briefly state the grounds upon which this opinion is founded.— Having been engaged for a considerable length of time in providing Food for the Poor at Munich, I was naturally led, as well by curiosity as motives of economy, to make a great variety of experiments upon that subject; and I had not proceeded far in my operations, before I began to perceive that they were very important;—even much more so than I had imagined.

The difference in the apparent goodness, of the palatableness, and apparent nutritiousness of the same kinds of Food, when prepared of cooked in different ways, struck me very forcibly; and I constantly found that the richness or QUALITY of a soup depended more upon a proper choice of the ingredients, and a proper management of the fire in the combination of those ingredients, than upon the quantity of solid nutritious matter employed;—much more upon the art and skill of the cook, than upon the amount of the sums laid out in the market.

I found likewise, that the nutritious of a soup, or its power of satisfying hunger, and affording nourishment, appeared always to be in proportion to its apparent richness or palatableness.

But what surprised me not a little, was the discovery of the very small quantity of SOLID FOOD, which, when properly prepared, will suffice to satisfy hunger, and support life and health; and the very trifling expence at which the stoutest, and most laborious man may, in any country, be fed.

After an experiment of more than five years in feeding the Poor at Munich during which time every experiment was made that could be devised, not only with regard to the choice of the articles used as

Food, but also in respect to their different combinations and proportions; and to the various ways in which they could be prepared or cooked; it was found that the CHEAPEST, most SAVOURY, and most NOURISHING Food that could be provided, was a soup composed of PEARL BARLEY, PEASE, POTATOES, CUTTINGS OF FINE WHEATEN BREAD, vinegar—salt and water in certain proportions.

The method of preparing this soup is as follows; The water and the pearl barley are first put together into the boiler and made to boil; the pease are then added, and the boiling is continued over a gentle fire about two hours;—the potatoes are then added, (having been previously peeled with a knife, or having been boiled, in order to their being more easily deprived of their skins,) and the boiling is continued for about one hour more, during which time the contents of the boiler are frequently stirred about with a large wooden spoon, or ladle, in order to destroy the texture of the potatoes, and to reduce the soup to one uniform mass.—When this is done, the vinegar and the salt are added; and last of all, at the moment it is to be served up, the cuttings of bread.

The soup should never be suffered to boil, or even to stand long before it is served up after the cuttings of bread are put into it. It will, indeed, for reasons which will hereafter be explained, be best never to put the cuttings of bread into the boiler at all, but, (as is always done at Munich,) to put them into the tubs in which the soup is carried from the kitchen into the dining-hall; pouring the soup hot from the boiler upon them; and stirring the whole well together with the iron ladles used for measuring out the soup to the Poor in the hall.

It is of more importance than can well be imagined, that this bread which is mixed with the soup should not be boiled. It is likewise of use that it should be cut as fine or thin as possible; and if it be dry and hard, it will be so much the better.

The bread we use at Munich is what is called semel bread, being small loaves, weighing from two to three ounces; and as we receive this bread in donations from the bakers, it is commonly dry and hard, being that which, not being sold in time, remains on hand, and becomes stale and unsaleable; and we have found by experience, that this hard and stale bread answers for our purpose much better than any other, for it renders mastication necessary; and mastication seems very powerfully to assist in promoting digestion: it likewise PROLONGS THE DURATION OF THE ENJOYMENT OF EATING, a matter of very great importance indeed, and which has not hitherto been sufficiently attended to.

The quantity of this soup furnished to each person, at each meal, or one portion of it, (the cuttings of bread included,) is just ONE BAVARIAN POUND in weight; and as the Bavarian pound is to the pound Avoirdupois as 1,123842 to 1, —it is equal to about nineteen ounces and nine-tenths Avoirdupois. Now, to those who know that a full pint of soup weighs no more than about sixteen ounces Avoirdupois, it will not, perhaps, at the first view, appear very extraordinary that a portion weighing near twenty ounces, and consequently making near ONE PINT AND A QUARTER of this rich, strong, savoury soup, should be found sufficient to satisfy the hunger of a grown person; but when the matter is examined narrowly, and properly analyzed, and it is found that the whole quantity of SOLID FOOD which enters into the composition of one of these portions of soup, does not amount to quite SIX OUNCES, it will then appear to be almost impossible that this allowance should be sufficient.

That it is quite sufficient, however, to make a good meal for a strong healthy person, has been abundantly proved by long experience. I have even found that a soup composed of nearly the same ingredients, except the potatoes, but in different proportions, was sufficiently nutritive, and very palatable, in which only about FOUR OUNCES AND THREE QUARTERS of solid Food entered into the composition of a portion weighing twenty ounces.

But this will not appear incredible to those who know, that one single spoonful of salope, weighing less than one quarter of an ounce, put into a pint of boiling water, forms the thickest and most nourishing soup that can be taken; and that the quantity of solid matter which enters into the composition of another very nutritive Food, hartshorn jelly, is not much more considerable.

The barley in my soup, seems to act much the same part as the salope in this famous restorative; and no substitute that I could ever find for it, among all the variety of corn and pulse of the growth of Europe, ever produced half the effect; that is to say, half the nourishment at the same expence. Barley may therefore be considered as the rice of Great Britain.

It requires, it is true, a great deal of boiling; but when it is properly managed, it thickens a vast quantity of water; and, as I suppose, PREPARES IT FOR DECOMPOSITION. It also gives the soup into which it enters as an ingredient, a degree of richness which nothing else can give. It has little or no taste in itself, but when mixed with other ingredients which are savoury, it renders them peculiarly grateful to the palate[2].

It is a maxim, as ancient, I believe, as the time of Hippocrates, that "whatever pleases the palate nourishes;" and I have often had reason to think it perfectly just. Could it be clearly ascertained and demonstrated, it would tend to place COOKERY in a much more respectable situation among the arts than it now holds.

That the manner in which Food is prepared is a matter of real importance; and that the water used in that process acts a much more important part than has hitherto been generally imagined, is, I think, quite evident; for, it seems to me to be impossible, upon any other suppositions, to account for the appearances. If the very small quantity of solid Food which enters into the composition of a portion of some very nutritive soup were to be prepared differently, and taken under some other form, that of bread, for instance; so far from being sufficient to satisfy hunger, and afford a comfortable and nutritive meal, a person would absolutely starve upon such a slender allowance; and no great relief would be derived from drinking CRUDE water to fill up the void in the stomach.

But it is not merely from an observation of the apparent effects of cookery upon those articles which are used as Food for man, that we are led to discover the importance of these culinary processes. Their utility is proved in a manner equally conclusive and satisfactory, by the efforts which have been produced by employing the same process in preparing Food for brute animals.

It is well known, that boiling the potatoes with which hogs are fed, renders them much more nutritive; and since the introduction of the new system of feeding horned cattle, that of keeping them confined in the stables all the year round, (a method which is now coming fast into common use in many parts of Germany,) great improvements have been made in the art of providing nourishment for those animals; and particularly by preparing their Food, by operations similar to those of cookery; and to these improvements it is most probably owing, that stall feeding has, in that country, been so universally successful.

It has long been a practice in Germany for those who fatten bullocks for the butcher, or feed milch-cows, to give them frequently what is called a drank or drink; which is a kind of pottage, prepared differently in different parts of the country, and in the different seasons, according to the greater facility with which one or other of the articles occasionally employed in the composition of it may be procured; and according to the particular fancies of individuals. Many feeders make a great secret of the composition of their drinks, and some have, to my knowledge, carried their refinement so far as actually to mix brandy in them, in small quantities; and pretend to have found their advantage in adding this costly ingredient.

The articles most commonly used are, bran, oatmeal, brewers grains, mashed potatoes, mashed turnips, rye meal, and barley meal, with a large proportion of water; sometimes two or three or more of these articles are united in forming a drink; and of whatever ingredients the drink is composed, a large proportion of salt is always added to it.

There is, perhaps, nothing new in this method of feeding cattle with liquid mixtures, but the manner in which these drinks are now prepared in Germany is, I believe, quite new; and shows what I wish to prove, that COOKING RENDERS FOOD REALLY MORE NUTRITIVE.

These drinks were formerly given cold, but it was afterwards discovered that they were more nourishing when given warm; and of late their preparation is, in many places, become a very regular culinary process. Kitchens have been built, and large boilers provided and fitted up, merely for cooking for the cattle in the stables; and I have been assured by many very intelligent farmers, who have adopted this new mode of feeding, (and have also found by my own experience,) that it is very advantageous indeed; that the drinks are evidently rendered much more nourishing and wholesome by being boiled; and that the expence of fuel, and the trouble attending this process, are amply compensated by the advantages derived from the improvement of the Food. We even find it advantageous to continue the boiling a considerable time, two or three hours, for instance; as the Food goes on to be still farther improved, the longer the boiling is continued[3].

These facts seem evidently to show, that there is some very important secret with regard to nutrition, which has not been yet properly investigated; and it seems to me to be more probable, that the numbers of inhabitants who may be supported in any country, upon its internal produce, depends almost as much upon the state of THE ART OF COOKERY, as upon that of agriculture. — The Chinese, perhaps, understand both these arts better than any other nation. — Savages understand neither of them.

But, if cookery be of so much importance, it certainly deserves to be studied with the greatest care; and it ought particularly to be attended to in times of general alarm on account of a scarcity of provisions; for the relief which may in such cases be derived from it, is immediate and effectual, while all other resources are distant and uncertain.

I am aware of the difficulties which always attend the introduction of measures calculated to produce and remarkable change in the customs and habits of mankind; and there is perhaps no change more difficult to effect, than that which would be necessary in order to make any considerable saving in the consumption of those articles commonly used as Food; but still, I am of opinion, that such a change might, with proper management, be brought about.

There was a time, no doubt, when an aversion to potatoes was as general, and as strong, in Great Britain, and even in Ireland, as it is now in some parts of Bavaria; but this prejudice has been got over; and I am persuaded, that any national prejudice, however deeply rooted, may be overcome, provided proper means be used for that purpose, and time allowed for their operation.

But notwithstanding the difficulty of introducing a general use of soups throughout the country, or of any other kind of Food, however palatable, cheap, and nourishing, to which people have not been accustomed, yet these improvements might certainly be made, with great facility, in all public hospitals and work-houses, where the Poor are fed at the public expense; and the saving of provisions, (not to mention the diminution of expence,) which might be derived from this improvement, would be very important at all times, and more especially in times of general scarcity.

Another measure, still more important, and which might, I am persuaded, be easily carried into execution, is the establishment of public kitchens in all towns, and large villages, throughout the kingdom, whence, not only the Poor might be fed gratis, but also all the industrious inhabitants of the neighbourhood might be furnished with Food at so cheap a rate, as to be a very great relief to them at all times; and in times of general scarcity, this arrangement would alone be sufficient to prevent those public and private calamities, which never fail to accompany that most dreadful of all visitations, a famine.

The saving of Food that would result from feeding a large proportion of the inhabitants of any country from public kitchens, would be immense, and that saving would tend, immediately, and most powerfully, to render provisions more plentiful and cheap,— diminish the general alarm on account of the danger of a scarcity, and prevent the hoarding up of provisions by individuals, which is often alone sufficient, without any thing else, to bring on a famine, even where there is no real scarcity: for it is not merely the FEARS of individuals which operate in these cases, and induce them to lay in a larger store of provisions than they otherwise would do; and which naturally increases the scarcity of provisions in the market, and raises their prices; but there are persons who are so lost to all the feelings of humanity, as often to speculate upon the distress of the Public, and all THEIR operations effectually tend to increase the scarcity in the markets, and augment the general alarm.

But without enlarging farther in this place upon these public kitchens, and the numerous and important advantages which may in all countries be derived from them, I shall return to the interesting subjects which I have undertaken to investigate;— the science of nutrition, and the art of providing wholesome and palatable Food at a small expence.

CHAPTER. II.

Of the Pleasure of Eating, and of the Means that may be employed for increasing it.

What has already been said upon this subject will, I flatter myself, be thought sufficient to show that, FOR ALL THE PURPOSES OF NOURISHMENT, a much smaller quantity of solid Food will suffice than has hitherto been thought necessary; but there is another circumstance to be taken into the account, and that is, the PLEASURE OF EATING;—an enjoyment of which no person will consent to be deprived.

The pleasure enjoyed in eating depends first upon the agreeableness of the taste of the Food; and secondly, upon its power to affect the palate. Now there are many substances extremely cheap, by which very agreeable tastes may be given to Food; particularly when the basis or nutritive substance of the Food is tasteless; and the effect of any kind of palatable solid Food, (of meat, for instance,) upon the organs of taste, may be increased, almost indefinitely, by reducing the size of the particles of such Food, and causing it to act upon the palate by a larger surface. And if means be used to prevent its being swallowed too soon, which may be easily done by mixing with it some hard and tasteless substance, such as crumbs of bread rendered hard by toasting, or any thing else of that kind, by which a long mastrication is rendered necessary, the enjoyment of eating may be greatly increased and prolonged.

The idea of occupying a person a great while, and affording him much pleasure at the same time, in eating a small quantity of Food, may, perhaps, appear ridiculous to some; but those who consider the matter attentively, will perceive that it is very important. It is, perhaps, as much so as any thing that can employ the attention of the philosopher.

The enjoyments which fall to the lot of the bulk of mankind are not so numerous as to render an attempt to increase them superfluous. And even in regard to those who have it in their power to gratify their appetites to the utmost extent of their wishes, it is surely rendering them a very importance service to show them how they may increase their pleasures without destroying their health.

If a glutton can be made to gormandize two hours upon two ounces of meat, it is certainly much better for him, than to give himself an indigestion by eating two pounds in the same time.

I was led to meditate upon this subject by mere accident. I had long been at a loss to understand how the Bavarian soldiers, who are uncommonly stout, strong, and healthy men, and who, in common with all other Germans, are remarkably fond of eating, could contrive to live upon the very small sums they expended for Food; but a more careful examination of the economy of their tables cleared up the point, and let me into a secret which awakened all my curiosity. These soldiers, instead of being starved upon their scanty allowance, as might have been suspected, I found actually living in a most comfortable and even luxurious manner. I found that they had contrived not only to render their Food savoury and nourishing, but, what appeared to me still more extraordinary, had found the means of increasing its action upon the organs of taste so as actually to augment, and even prolong to a most surprising degree, the enjoyment of eating.

This accidental discovery made a deep impression upon my mind, and gave a new turn to all my ideas on the subject of Food. — It opened to me a new and very interesting field for investigation and experimenting inquiry, of which I had never before had a distinct view; and thenceforward my diligence in making experiments, and in collecting information relative to the manner in which Food is prepared in different countries, was redoubled.

In the following Chapter may be seen the general results of all my experiments and inquiries relative to this subject. — A desire to render this account as concise and short as possible has induced me to omit much interesting speculation which the subject naturally suggested; but the ingenuity of the reader will supply this defect, and enable him to discover the objects particularly aimed at in the experiments, even where they are not mentioned, and to compare the results of practice with the assumed theory.

CHAPTER. III.

Of the different kinds of food furnished to the poor in the
  house of industry at Munich, with an account of the cost of them.
Of the Expense of providing the same kinds of food in Great
  Britain, as well at the present high prices of provisions,
  as at the ordinary prices of them.
Of the various improvements of which these different kinds of
  cheap food are capable.

Before the introduction of potatoes as Food in the House of Industry at Munich, (which was not done till last August,) the Poor
were fed with a soup composed in the following manner:

SOUP No I. Weight Cost in Ingredients Avoirdupois sterling
money. lb. oz. L. s. d. 4 viertls[4] of pearl barley, equal to about 20
1/3 gallons … … … 141 2 0 11 7 1/2 4 viertls of peas … … … … 131
4 0 7 3 1/4 Cuttings of fine wheaten bread … 69 10 0 10 2 1/4 Salt …
… … … … … … 19 13 0 1 2 1/2 24 maass, very weak beer — vinegar,
or rather small beer turned sour, about 24 quarts … … … … … … 46
13 0 1 5 1/2 Water, about 560 quarts … … … 1077 0 — — — — — — —
— — — - 1485 10 1 11 8 13/22

Brought over 1 11 8 13/22 Fuel, 88lb. of dry pine wood, the Bavarian clafter, (weighing 3961 lb. avoirdupois,) at 8s. 2 1/4d. sterling[5]
… … … … … 0 0 2 1/4 Wages of three cook-maids, at twenty florins
(37s. 7 1/2d.) a year, makes daily … … … 0 0 3 2/3 Daily expence
for feeding the three cook-maids, at ten creutzers (3 2/3 pence sterling) each, according to an agreement made with them … … 0 0 11
Daily wages of two men servants, employed in going to market —
collecting donations of bread, etc. helping in the kitchen, and assisting in serving out the soup to the Poor … … … 0 1 7 1/4 Repairs of
the kitchen, and of the kitchen furniture, about 90 florins (8L. 3s. 7d.
sterling) a year, makes daily … … … … … … 0 0 5 1/2 — — — — —
— - Total daily expense, when dinner is provided for 1200 persons
… … … … … … … 1 15 2 1/4

This sum (1L. 15s. 2 1/4d.) divided by 1200, the number of portions of soup furnished, gives for each portion a mere trifle more than ONE THIRD OF A PENNY, or exactly 422/1200 of a penny; the weight of each portion being about 20 ounces.

But, moderate as these expenses are, which have attended the feeding of the Poor of Munich, they have lately been reduced still farther by introducing the use of potatoes. — These most valuable vegetables were hardly known in Bavaria till very lately; and so strong was the aversion of the public, and particularly of the Poor, against them, at the time when we began to make use of them in the public kitchen of the House of Industry in Munich, that we were absolutely obliged, at first, to introduce them by stealth. — A private room in a retired corner was fitted up as a kitchen for cooking them; and it was necessary to disguise them, by boiling them down entirely, and destroying their form and texture, to prevent their being detected: — but the Poor soon found that their soup was improved in its qualities; and they testified their approbation of the change that had been made in it so generally and loudly, that it was at last thought to be no longer necessary to conceal from them the secret of its composition, and they are now grown so fond of potatoes that they would not easily be satisfied without them.

The employing of potatoes as an ingredient in the soup has enabled us to make a considerable saving in the other more costly materials, as may be seen by comparing the following receipt with that already given.

SOUP, No II.

Ingredients. Weight Cost in Avoirdupois. sterling money. lb. oz. L. s. d. 2 viertls of pearl barley … … 70 9 0 5 9 13/22 2 viertls of peas … … … 65 10 0 3 7 5/8 8 viertls of potatoes … … 230 4 0 1 9 9/11 Cuttings of bread … … … 69 10 0 10 2 4/11 Salt … … … … … … 19 13 0 1 2 1/2 Vinegar … … … … … 46 13 0 1 5 1/2 Water … … … … … … 982 15 − − − −- Total weight 1485 10 Expenses for fuel, servants, repairs, etc. as before … … … … … … … 0 3 5 5/12 − − − − − − − Total daily expence, when dinner is provided for 1200 persons … … … … … … … 1 7 6 2/3

This sum (1L. 7s. 6 2/3.) divided by 1200, the number of portions of soup, gives for each portion ONE FARTHING very nearly; or accurately, 1 1/40 farthing.

The quantity of each of the ingredients contained in one portion of soup is as follows:

In avoirdupois weight.
Ingredients. Soup, No I. Soup, No II.

Of pearl barley 1 1058/1200 0 1129/1200
Of peas … … 1 960/1200 0 1050/1200
Of potatoes … – – – 3 84/1200
Of bread … … 0 1114/1200 0 1114/1200
—————–·—————–
Total solids 4 772/1200 5 977/1200
Of salt … … 0 316/1200 0 316/1200
Of weak vinegar 0 748/1200 0 748/1200
Of water … … 14 432/1200 13 127/1200
—————–·—————–
Total 19 968/1200 19 968/1200

The expence of preparing these soups will vary with the prices of the articles of which they are composed; but as the quantities of the ingredients, determined by weight, are here given, it will be easy to ascertain exactly what they will cost in any case whatever.

Suppose, for instance, it were required to determine how much 1200 portions of the Soup, No. I. would cost in London at this present moment, (the 12th of November 1795,) when all kinds of provisions are uncommonly dear. I see by a printed report of the Board of Agriculture, of the day before yesterday (November 10), that the prices of the articles necessary for preparing these soups were as follows:

Barley, per bushel weighing 46lb. at 5s. 6d. which gives for each pound about 1 1/2d; but prepared as pearl barley, it will cost at least two pence per pound[6].

Boiling peas per bushel, weighing 61 1/4lb. (at 10s.) which gives for each pound nearly 1 1/2d.

Potatoes, per bushel, weighing 58 1/2lb. at 2s. 6d. which gives nearly one halfpenny for each pound.

And I find that a quartern loaf of wheaten bread, weighing 4lb. 5oz. costs now in London 1s. 0 1/4d.;—this bread must therefore be reckoned at 11 25/69 farthings per pound.

Salt costs 1 1/2. per pound; and vinegar (which is probably six times as strong as that stuff called vinegar which is used in the kitchen of the House of Industry at Munich) costs 1s. 8d. per gallon.

This being premised, the computations may be made as follows:

Expence of preparing in London, in the month of November 1795, 1200 portions of the Soup, No I.

lb oz s d L. s. d. 141 2 pearl barley, at 0 2 per lb. 1 12 6 131 4 peas, at 0 1 1/2 — — — 0 16 4 69 10 wheaten bread, at 0 11 25/99 — — — 0 16 6 19 13 salt, at 0 1 1/2 — — — 0 2 5 1/2 Vinegar, one gallon, at 1 8 — — — 0 1 8 Expences for fuel, servants, kitchen furniture, etc. reckoning three times as much as those articles of expence amount to daily at Munich … … … … … 0 10 4 1/4 — — — — — — —- Total 3 9 9 1/4

Which sum (3L. 9s. 9 1/4d.) divided by 1200, the number of portions of soup, gives 2 951/1200 farthings, or nearly 2 3/4 farthings for each portion.

For the Soup, No II. it will be,
lb. oz. s. d. L. s. d.
70 9 pearl barley, at 0 2 — — — 0 11 9
65 10 peas, at 0 1 1/2 — — — 0 8 2
230 4 potatoes, at 0 0 1/2 — — — 0 13 9
69 10 bread, at 0 11 25/65 — — — 0 16 6
19 13 salt, at 0 1 1/2 — — — 0 2 5 1/2
Vinegar, one gallon — — — 0 1 8
Expenses for fuel, servants, etc. — — — 0 10 4 1/4
— — — — — — —-
Total 3 4 7 3/4

This sum (3L. 4s. 7 3/4d.) divided by 1200, the number of portions, gives for each 2 1/2 farthings very nearly.

This soup comes much higher here in London, than it would do in most other parts of Great Britain, on account of the very high price of potatoes in this city; but in most parts of the kingdom, and certainly in every part of Ireland, it may be furnished, even at this present moment, notwithstanding the uncommonly high prices of provisions, at less than ONE HALFPENNY the portion of 20 ounces.

Though the object most attended to in composing these soups was to render them wholesome and nourishing, yet they are very far from being unpalatable. — The basis of the soups, which is water prepared and thickened by barley, is well calculated to receive, and to convey to the palate in an agreeable manner, every thing that is savoury in the other ingredients; and the dry bread rendering mastication necessary, prolongs the action of the Food upon the organs of taste, and by that means increases and PROLONGS the enjoyment of eating.

But though these soups are very good and nourishing, yet they certainly are capable of a variety of improvements. — The most obvious means of improving them is to mix with them a small quantity of salted meat, boiled, and cut into very small pieces, (the smaller the better,) and to fry the bread that is put into them in butter, or in the fat of salted pork or bacon.

The bread, by being fried, is not only rendered much harder, but being impregnated with a fat or oily substance it remains hard after it is put into the soup, the water not being able to penetrate it and soften it.

All good cooks put fried bread, cut into small square pieces, in peas-soup; but I much doubt whether they are aware of the very great importance of that practice, or that they have any just idea of the MANNER in which the bread improves the soup.

The best kind of meat for mixing with these soups is salted pork, or bacon, or smoked beef.

Whatever meat is used, it ought to be boiled either in clear water or in the soup; and after it is boiled, it ought to be cut into very small pieces, as small perhaps, as barley-corns. — The bread may be cut in pieces of the size of large peas, or in thin slices; and after it is fried, it may be mixed with the meat and put into the soup-dishes, and the soup poured on them when it is served out.

Another method of improving this soup is to mix it with small dumplins, or meat-balls, made of bread, flour, and smoked beef, ham, or any other kind of salted meat, or of liver cut into small pieces, or rather MINCED, as it is called. — These dumplins may be boiled either in the soup or in clear water, and put into the soup when it is served out.

As the meat in these compositions is designed rather to please the palate than for any thing else, the soup being sufficiently nourishing without it, it is or much importance that it be reduced to very small pieces, in order that it be brought into contract with the organs of taste by a large surface; and that it be mixed with some hard substance, (fried bread, for instance, crumbs, or hard dumplins,) which will necessarily prolong the time employed in mastication.

When this is done, and where the meat employed has much flavour, a very small quantity of it will be found sufficient to answer the purpose required.

ONE OUNCE of bacon, or of smoked beef, and ONE OUNCE of fried bread, added to EIGHTEEN OUNCES of the Soup No. I. would afford an excellent meal, in which the taste of animal food would decidedly predominate.

Dried salt fish, or smoked fish, boiled and then minced, and made into dumplins with mashed potatoes, bread, and flour, and boiled again, would be very good, eaten with either of the Soup No. I. or No. II.

These soups may likewise be improved, by mixing with them various kinds of cheap roots and green vegetables, as turnips, carrots, parsnips, celery, cabbages, sour-crout, etc. as also by seasoning them with fine herbs and black pepper. — Onions and leeks may likewise be used with great advantage, as they not only serve to render the Food in which they enter as ingredients peculiarly savoury, but are really very wholesome.

With regard to the barley made use of in preparing these soups, though I always have used pearl barley, or rolled barley(as it is called in Germany), yet I have no doubt but common barley-meal would answer nearly as well; particularly if care were taken to boil it gently for a sufficient length of time over a slow fire before the peas are added[7].

Till the last year, we used to cook the barley-soup and the peas-soup separate, and not to mix them till the moment when they were poured into the tubs upon the cut bread, in order to be carried into the dining-hall; but I do not know that any advantages were derived from that practice; the soup being, to all appearances, quite as good since the barley and the peas have been cooked together as before.

As soon as the soup is done, and the boilers are emptied, they are immediately refilled with water, and the barley for the soup for the next day is put into it, and left to steep over night; and at six o'clock the next morning the fires are lighted under the boilers[8].

The peas, however, are never suffered to remain in the water over-night, as we have found, by repeated trials, that they never boil soft if the water in which they are boiled is not boiling hot when they are put into it. — Whether this is peculiar to the peas which grow in Bavaria, I know not.

When I began to feed the Poor of Munich, there was also a quantity of meat boiled in their soup; but as the quantity was small, and the quality of it but very indifferent, I never thought it contributed much to rendering the victuals more nourishing: but as soon as means were found for rendering the soup palatable without meat, the quantity of it used was gradually diminished, and it was at length entirely omitted. I never heard that the Poor complained of the want of it; and much doubt whether they took notice of it.

The management of the fire in cooking is, in all cases, a matter of great importance; but in no case is it so necessary to be attended to as in preparing the cheap and nutritive soups here recommended. — Not only the palatableness, but even the strength or richness of the soup, seems to depend very much upon the management of the heat employed in cooking it.

From the beginning of the process to the end of it, the boiling should be as gentle as possible; — and if it were possible to keep the soup always JUST BOILING HOT, without actually boiling, it would be so much the better.

Causing any thing to boil violently in any culinary process is very ill judged; for it not only does not expedite, even in the smallest degree, the process of cooking, but it occasions a most enormous waste of fuel; and by driving away with the steam many of the more volatile and more savoury particles of the ingredients, renders the victuals less good and less palatable. —To those who are acquainted with the experimental philosophy of heat, and who know that water once brought to be BOILING HOT, however gently it may boil in fact, CANNOT BE MADE ANY HOTTER, however large and intense the fire under it may be made, and who know that it is by the HEAT—that is to say, THE DEGREE or intensify of it, and the TIME of its being continued, and not by the bubbling up or BOILING, (as it is called) of the water that culinary operations are performed—this will be evident, and those who know that more than FIVE TIMES as much heat is required to SEND OFF IN STEAM any given quantity of water ALREADY BOILING HOT as would be necessary to heat the same quantity of ICE-COLD water TO THE BOILING POINT —will see the enormous waste of heat, and consequently of fuel, which, in all cases must result from violent boiling in culinary processes.

To prevent the soup from burning to the boiler, the bottom of the boiler should be made DOUBLE; the false bottom, (which may be very thin) being fixed on the inside of the boiler, the two sheets of copper being every where in contact with each other; but they ought not to be attached to each other with solder, except only at the edge of the false bottom where it is joined to the sides of the boiler.—The false bottom should have a rim about an inch and a half wide, projecting upwards, by which it should be riveted to the sides of the boiler; but only few rivets, or nails, should be used for fixing the two bottoms together below, and those used should be very small; otherwise where large nails are employed at the bottom of the boiler, where the fire is most intense, the soup will be apt to BURN TO; at least on the heads of those large nails.

The two sheets of metal may be made to touch each other every where, by hammering them together after the false bottom is fixed in its place; and they may be tacked together by a few small rivets placed here and there, at considerable distances from each other; and after this is done, the boiler may be tinned.

In tinning the boiler, if proper care be taken, the edge of the false bottom may be soldered by the tin to the sides of the boiler, and this will prevent the water, or other liquids put into the boiler, from getting between the two bottoms.

In this manner double bottoms may be made to sauce-pans and kettles of all kinds used in cooking; and this contrivance will, in all cases, most effectually prevent what is called by the cooks burning to[9].

The heat is so much obstructed in its passage through the thin sheet of air, which, notwithstanding all the care that is taken to bring the two bottoms into actual contact, will still remain between them, the second has time to give its heat as fast as it receives it, to the fluid in the boiler; and consequently never acquires a degree of heat sufficient for burning any thing that may be upon it.

Perhaps it would be best to double copper sauce-pans and small kettles throughout; and as this may and ought to be done with a very thin sheet of metal, it could not cost much, even if this lining were to be made of silver.

But I must not enlarge here upon a subject I shall have occasion to treat more fully in another place. — To return, therefore, to the subject more immediately under consideration, Food.

CHAPTER. IV.

Of the small expense at which the Bavarian soldiers are fed.
Details of their housekeeping, founded on actual experiment.
An account of the fuel expended by them in cooking.

It has often been matter of surprise to many, and even to those who are most conversant in military affairs, that soldiers can find means to live upon the very small allowances granted them for their subsistence; and I have often wondered that nobody has undertaken to investigate that matter, and to explain a mystery at the same time curious and interesting, in a high degree.

The pay of a private soldier is in all countries very small, much less than the wages of a day-labourer; and in some countries it is so mere a pittance, that it is quite astonishing how it can be made to support life.

The pay of a private foot-soldier in the service of His Most Serene Highness the Elector Palatine, (and it is the same for a private grenadier in the regiment of guards,) is FIVE CREUTZERS a-day, and no more.—Formerly the pay of a private foot-soldier was only four creutzers and a half a-day, but lately, upon the introduction of the new military arrangements in the country, his pay has been raised to five creutzers;—and with this he receives one pound thirteen ounces and a half, Avoirdupois weight, of rye-bread, which, at the medium price of grain in Bavaria and the Palatinate, costs something less than three creutzers, or just about ONE PENNY sterling.

The pay which the soldier receives in money,— (five creutzers a-day,) equal to one penny three farthings sterling, added to his daily allowance of bread, valued at one penny, make TWO PENCE THREE FARTHINGS a-day, for the sum total of his allowance.

That it is possible, in any country, to procure Food sufficient to support life with so small a sum, will doubtless appear extraordinary to an English reader;—but what would be his surprise upon seeing a whole army, composed of the finest, stoutest, and strongest men in the world, who are fed upon that allowance, and whose countenances show the most evident marks of ruddy health, and perfect contentment?

I have already observed, how much I was struck with the domestic economy of the Bavarian soldiers. I think the subject much too interesting, not to be laid before the Public, even in all its details; and as I think it will be more satisfactory to hear from their own mouths an account of the manner in which these soldiers live, I shall transcribe the reports of two sensible non-commissioned officers, whom I employed to give me the information I wanted.

These non-commissioned officers, who belong to two different regiments of grenadiers in garrison at Munich, were recommended to me by their colonels as being very steady, careful men, are each at the head of a mess consisting of twelve soldiers, themselves reckoned in the number. The following accounts, which they gave me of their housekeeping, and of the expenses of their tables, were all the genuine results of actual experiments made at my particular desire, and at my cost.

I do not believe that useful information was ever purchased cheaper than upon this occasion; and I fancy my reader will be of the same opinion when he has perused the following reports, which are literally translated from the original German.

"In obedience to the orders of Lieut. General Count Rumford, the following experiments were made by Serjeant Wickenhof's mess, in the first company of the first (or Elector's own) regiment of grenadiers, at Munich, on the 10th and 11th of June 1795.

June 10th, 1795.
BILL OF FARE
Boiled beef, with soup and bread dumplins.
Details of the expence, etc.
For the boiled beef and the soup.

    lb. loths. Creutzers.
    2 0 beef[10] 16
    0 1 sweet herbs 1
    0 0 1/2 pepper 0 1/2
    0 6 salt 0 1/2
    1 14 1/2 ammunition bread, cut fine 2 7/8

9 20 water 0
— — —- — — — —
Total 13 10 Cost 20 7/8

All these articles were put together into an earthen pot, and boiled two hours and a quarter. The meat was then taken out of the soup and weighed, and found to weigh 1 lb. 30 loths; which, divided into twelve equal portions, gave FIVE LOTHS for the weight of each.

The soup, with the bread, etc. weighed 9 lb. 30 1/2 loths; which, divided into twelve equal portions, gave for each 26 7/12 loths.

The cost of the meat and soup together, 20 7/8 creutzers, divided by twelve, gives 1 3/4 creutzers, very nearly, for the cost of each portion.

For the bread dumplins.

lb. loths. Creutzers.
1 13 of fine semel bread 10
1 0 of fine flour ... 4 1/2
0 6 salt 0 1/2
3 0 of water 0
— — — — — — —
Total 5 19 Cost 15

This mass was made into dumplins, and these dumplins were boiled half an hour in clear water. Upon taking them out of the water, they were found to weigh 5 lb. 24 loths; and dividing them into twelve equal portions, each portion weighed 15 1/3 loths; and the cost of the whole (15 creutzers), divided by twelve, gives 1 1/4 creutzers for the cost of each portion.

The meat, soup, and dumplins were served all at once in the same dish, and were all eaten together; and with this meal, (which was their dinner, and was eat at twelve o'clock,) each person belonging to the mess was furnished with a piece of rye-bread, weighing ten loths, and which cost 5/16 of a creutzer. —Each person was likewise furnished with a piece of this bread, weighing ten loths, for his breakfast;—another piece, of equal weight, in the afternoon at four o'clock; and another in the evening.

Analysis of this Day's Fare.

Each person received in the Amount of cost in course of the day Bavarian money.

In solids. In fluids.
        lb. loths. lb. loths. Creutzers.
Boiled beef 0 5 ... ... 1 1/6
In the soup.
  Rye-bread 0 3 7/8]
  Sweet herbs 0 0 1/12]
  Salt ... ... 0 0 1/24].... 0 7/16
  Pepper ... ... 0 0 1/24]
  Water ... ... 0 23 1/2]
         — — — — — — — — —-]
     Total 0 4 2/24 0 23 1/2]

In dumplins.
  Wheaten-bread 0 3 3/4]
  Ditto flour 0 2 2/3]
  Salt ... ... 0 0 1/24].... 1 1/4
  Water ... ... 0 7 1/12]
         — — — — — — — — —-]
     Total 0 6 11/24 0 7 7/12]

Dry bread.
  For breakfast 0 10]
  At dinner 0 10]
  In the afternoon 0 10].... 2 1/2
  At supper 0 10]
        — — —]
    Total 1 8]
       — — — — — — — —
General total 2 24 13/24 0 31 1/2 which cost 5 17/48

The ammunition bread is reckoned in this estimate at two creutzers the Bavarian pound, which is about what it costs at a medium;

and as the daily allowance of the soldiers is 1 1/2 Bavarian pounds of the bread, this reckoned in money amounts to three creutzers a-day; and this added to his pay at five creutzers a-day, makes eight creutzers a-day, which is the whole of his allowance from the sovereign for his subsistence.

But it appears from the foregoing account, that he expends for Food no more than 5 17/48 creutzers a-day, there is therefore a surplus amounting to 2 31/48 creutzers a-day, or very near ONE-THIRD OF HIS WHOLE ALLOWANCE, which remains; and which he can dispose of just as he thinks proper.

This surplus is commonly employed in purchasing beer, brandy, tobacco, etc. Beer in Bavaria costs two creutzers a pint, brandy, or rather malt-spirits, from fifteen to eighteen creutzers; and tobacco is very cheap.

To enable the English reader to form, without the trouble of computation, a complete and satisfactory idea of the manner in which these Bavarian soldiers are fed, I have added the following Analysis of their fare; in which the quantity of each article is expressed in Avoirdupois weight, and its cost in English money.

Analysis.

Each person belonging to the mess received in the course of the day, Cost in English June 11th, 1795. money.

lb. oz. s. d.
Dry ammunition bread 1 8 76/100 0 0 10/11
Ammunition bread cooked
  in the soup 0 2 4/10 0 0 23/264
Fine wheaten (semel)
  bread in the dumplins ... 0 2 3/10 0 0 10/33

           — — — — —

           Total bread 1 13 46/100

Fine flour in the dumplins 0 1 65/100 0 0 18/33
Boiled beef 0 3 1/10 0 0 72/198
In seasoning; fine herbs,
  salt and pepper 0 0 13/100 0 0 2/33

           — — — — — —-

Total solids 2 2 34/100

Water prepared by cooking. In the soup ... ... 0 14 52/209 In the dumplins ... ... 0 4 32/100 — — — — —- Total prepared water 1 2 84/100 — — — — —- Total solids and fluids 3 5 18/100

Total expense for each person 5 17/48 creutzers, equal to TWO PENCE sterling, very nearly.

But as the Bavarian soldiers have not the same fare every day, the expences of their tables cannot be ascertained from one single experiment. I shall therefore return to Serjeant Wickenhof's report.

11th of June 1795.
Bill of Fare.
Bread, dumplins, and soup.
Details of expenses, etc.

      For the dumplins.
lb. loths. Creutzers.
2 13 wheaten bread 14
0 16 butter 9
1 0 fine flour 4 1/2
0 11 eggs 3
0 6 salt 0 1/2
0 0 1/2 pepper 0 1/2
3 16 water ... ...
— — —- — — —-
7 30 1/2 Cost 31 1/2 creutzers.

This made into dumplins;—the dumplins, after being boiled, were found to weigh eight pounds eight loths, which, divided among twelve persons, gave for each twenty-two loths.—And the cost of the whole (31 1/2 creutzers), divided by 12, gives 2 15/24 creutzers for each portion.

      For the soup.
lb. loths. Creutzers.
1 14 1/2 ammunition bread ... 2 7/8
0 6 salt 0 1/2

0 1 sweet herbs 1
12 0 water ... ...
———— · ————·
13 21 1/2 Cost 4 3/8 creutzers.

This soup, when cooked, weighed 11 lb, 26 loths; which, divided among the twelve persons belonging to the mess, gave for each 31 1/2 loths; and the cost (4 3/8 creutzers), divided by twelve, gives nearly THREE-NINTHS of a creutzer for each portion.

For bread.

Four pieces of ammunition bread, weighing each ten loths, for each person, — namely, one piece for breakfast — one at dinner — one in the afternoon, — and one at supper; in all, 40 loths, or one pound and a quarter, costs two creutzers and a half.

Details of expenses, etc. for each person.

lb. loths. Creutzers
For 1 8 dry bread 2 1/2
For 0 22 bread dumplins ... 2 15/24
For 0 31 1/2 bread soup 0 3/8
————— · ——·
2 30 1/2 of Food Cost 5 1/2 creutzers.

The same details expressed in Avoirdupois weight, and English money:

For each person
 lb. oz. Pence
 1 8 76/100 dry ammunition bread 0 10/11
 0 13 6/10 bread dumplins ... 0 693/792
 1 3 1/2 bread soup 0 36/264
————— ————·
3 9 86/100 of Food Cost 2 pence.

June 20th, 1795.
Serjeant Kein's mess, second regiment of grenadiers.

   Bill of Fare.

Boiled beef—bread soup—and liver dumplins.
Details of expenses, etc.
For the boiled beef and soup.

lb. loths. Creutzers.
 2 0 beef 15
 0 6 1/2 salt 0 1/2
 0 0 1/2 pepper 0 1/2
 0 2 sweet herbs ... 0 1/2
 2 24 ammunition bread 3 1/4
17 0 water... ... ...
 — — — — — — — — —-
22 1 Cost 19 1/2 creutzers.

These ingredients were all boiled together two hours and five minutes; after which the beef was taken out of the soup and weighed, and was found to weigh 1 lb. 22 loths; the soup weighed 15 lb.; and these divided equally among the twelve persons belonging to the mess, gave for each portion, 4 1/2 loths of beef, and 1 lb. 8 loths of soup; and the cost of the whole (19 3/4 creutzers), divided by 18, gives 1 31/48 creutzers for the cost of each portion.

Details of expenses, etc. for the liver dumplins.

lb. loths. Creutzers.
 2 28 of fine semel bread 15
 1 0 of beef liver 5
 0 18 of fine flour 2 1/2
 0 6 of salt 0 1/2
 2 24 of water ... ... —-
 — — — — — — — — —
Total 7 12 Cost 23 creutzers.

These ingredients being made into dumplins, the dumplins after being properly boiled were found to weigh 8 lb. — This gave for each portion 21 1/3 loths; and the amount of the cost (23 creutzers), divided by 12, the number of the portions, gives for each 1 11/12 creutzers.

The quantity of dry ammunition bread furnished to each person was 1 lb. 8 loths; and this, at two creutzers a pound, amounts to 2 1/2 creutzers.

Recapitulation.

For each person
  lb. loths. Creutzers.
   0 4 1/2 of boiled beef, and] … 1 31/48
   1 8 of bread soup]
   0 21 1/4 of liver dumplins … … 1 11/12
   1 8 of dry bread … … … 2 1/2
  — — — — — — — — — — — —-

  3 9 5/6 of Food Cost 6 3/48 creutzers.

In Avoirdupois weight, and English money, it is, — for each person:

  lb. oz.
   0 2.78 of boiled beef, and] … 0 948/1584
   1 8.91 of bread soup]
   0 13.19 of liver dumplins … … 0 276/306
   1 8.76 of dry bread … … … 0 10/11
  — — — —- — — — — — —-

  4 1.54 of Food Cost 2 1/5 pence.

June 21st, 1795.
Bill of Fare.
Boiled beef, and bread soup, with bread dumplins.
Details of expenses, etc. for the boiled beef and bread soup.
The same as yesterday,
For the dumplins.

lb. loths. Creutzers. 2 30 semel bread ... ... 15 1/2 0 18 fine flour ... ... 3 0 6 salt ... ... 0 1/2 3 0 water ... ... — — —- — — — —- 6 22 Cost 19 creutzers.

These dumplins being boiled, were found to weigh 7 lb. which gave for each person 18 2/3 loths; and each portion cost 1 7/12 creutzers.

Dry ammunition bread furnished to each person 1 lb. 8 loths, which cost 2 1/2 creutzers.

Recapitulation.

Each person belonging to the mess received this day:

lb. loths. Creutzers. 0 4 1/2 of boiled beef, and] ... 1 31/48 1 8 of bread soup] 0 18 2/3 of bread dumplins ... ... 1 7/12 1 8 of dry bread ... ... 2 1/2 — — — — —- — — — —- 3 7 1/6 of Food Cost 5 35/42 creutzers

In Avoirdupois weight, and English money, it is,

lb. oz.
0 2.78 of boiled beef, and] ... 0 948/1584
1 8.76 of bread soup]
0 11.54 of bread dumplins 0 228/396
1 8.76 of dry bread 0 10/11
— — —- — — — — —
4 0 of Food Cost 2 1/12 pence.

June 22d, 1795.
Bill of Fare.
Bread soup and meat dumplins.
Details of expenses, etc.

lb. loths.
2 0 of beef 15
2 30 of semel bread ... 15 1/2
0 18 of fine flour 3
0 1 of pepper 1
0 12 of salt 1
0 2 of sweet herbs ... 0 1/2
2 24 of ammunition bread 3 1/4

2 16 of water to the dumplins

— — —

Cost 39 1/4 creutzers.

The meat being cut fine, or minced, was mixed with the semel or wheaten bread; and these with the flour, and a due proportion of salt, were made into dumplins, and boiled in the soup.—These dumplins when boiled, weighed 10 lb. which, divided into 12 equal portions, gave 20 2/3 loths for each.

The soup weighed 15 lb. which gave 1 lb. 8 loths for each portion. —Of dry ammunition bread, each person received 1 lb. 8 loths, which cost 2 1/2 creutzers.

Recapitulation.

Each person received this day

lb. loths. Creutzers 0 20 2/3 of meat dumplins, and] ... 3 13/48 1 8 of bread soup] 1 8 of ammunition bread 2 1/2 — — — — —· — — —-- 3 4 2/3 of Food Cost 5 37/48 creutzers.

In Avoirdupois weight, and English money, it is,

lb. oz. Pence. 0 12.77 of meat dumplins, and] ... 1 300/1584 1 8.76 of bread soup] 1 8.76 of ammunition bread ... ... 0 10/11 — — — — — — — — — — 3 14.29 of Food Cost 2 1/10 pence.

The results of all these experiments, (and of many more which I could add,) show that the Bavarian soldier can live,—and the fact is that he actually does live,—upon a little more than TWO THIRDS of his allowance.—Of the five creutzers a-day which he receives in money, he seldom puts more than two creutzers and a half, and never more than three creutzers into the mess; so that at least TWO-FIFTHS of his pay remains, after he has defrayed all the expenses of his subsistence; and as he is furnished with every article of his clothing by the sovereign, and no stoppage is ever permitted to be made of any part of his pay, on any pretence whatever, THERE IS NO SOLDIER IN EUROPE WHOSE SITUATION IS MORE COM-FORTABLE.

Though the ammunition bread with which he is furnished is rather coarse and brown, being made of rye-meal, with only a small quantity of the coarser part of the bran separated from it, yet it is not only wholesome, but very nourishing; and for making soup it is even more palatable than wheaten bread. Most of the soldiers, however, in the Elector's service, and particularly those belonging to the Bavarian regiments, make a practice of selling a great part of their allowance of ammunition bread, and with the money they get for it, buy the best wheaten bread that is to be had; and many of them never taste brown bread but in their soup.

The ammunition bread is delivered to the soldiers every fourth day, in loaves, each loaf being equal to two rations; and it is a rule generally established in the messes, for each soldier to furnish one loaf for the use of the mess every twelfth day, so that he has five-sixths of his allowance of bread, which remains at his disposal.

The foregoing account of the manner in which the Bavarian soldiers are fed, will, I think, show most clearly the great importance of making soldiers live together in messes. — It may likewise furnish some useful hints to those who may be engaged in feeding the Poor, or in providing Food for ships's companies, or other bodies of men who are fed in common.

With regard to the expense of fuel in these experiments, as the victuals were cooked in earthen pots, over an open fire, the consumption of fire-wood was very great.

On the 10th of June, when 9 lb. 30 1/2 loths of soup, 1 lb. 28 loths of meat, and 5 lb. 24 loths of bread dumplins, in all 17 lb. 18 1/2 of Food were prepared, and the process of cooking, from the time the fire was lighted till the victuals were done, lasted two hours and forty-five minutes, and twenty-nine pounds, Bavarian weight, of fire-wood were consumed.

On the 11th of June, when 11 lb. 26 loths of bread soup, and 8 lb. 8 loths of bread dumplins, in all 20 lb. 2 loths of Food were prepared, the process of cooking lasted one hour and thirty minutes; — and seventeen pounds of wood were consumed.

On the 20th of June, in Serjeant Kein's mess, 15 lb. of soup; 1 lb. 22 loths of meat, and 8 lb. of liver dumplins, in all 24 lb. 22 loths of Food were prepared, and through the process of cooking lasted two hours and forty-five minutes, only 27 1/2 lb. of fire-wood were consumed.

On the 21st of June, the same quantity of soup and meat, and 7 lb. of bread dumplins, in all 23 lb. 22 loths of Food were prepared in two hours and thirty minutes, with the consumption of 18 1/2 lb. of wood.

On the 22nd of June, 15 lb. of soup, and 10 lb. of meat dumplins, in all 25 lb. of Food, were cooked in two hours and forty-five minutes, and the wood consumed was 18 lb. 10 loths.

The following table will show, in a striking and satisfactory manner, the expense of fuel in these experiments:

Date of the Time employed Quantity Quantity Quantity Experiments. in cooking. of Food of Wood of Wood to prepared. consumed. 1 lb. of Food.

June 1795. Hours. min. lb. loths. lb.
   10th 2 45 17 18 1/2 29
   11th 1 30 20 2 17
   20th 2 45 24 22 17 1/2
   21st 2 30 23 22 18 1/2
   22d 2 45 25 0 18 1/4

     — — — — — — — — — —. — — —-

Sums 5 12 15 111 0 1/2 100 1/4

     — — — — — — — — — —. — — —-

Means 2 23 22 0 1/5 20 1/20 10/11 lb.

The mean quantity of Food prepared daily in five days being 22 lb. very nearly, and the mean quantity of fire-wood consumed being 20 1/20 lb.; this gives 10/11 lb. of wood for each pound of Food.

But it has been found by actual experiment, made with the utmost care, in the new kitchen of the House of Industry at Munich, and often repeated, that 600 lb. of Food, (of the Soup No. I. given to the Poor,) may be cooked with the consumption of only 44 lb. of pine-wood. And hence it appears how very great the waste of fuel must be in all culinary processes, as they are commonly performed; for though the time taken up in cooking the soup for the Poor is, at a medium, more than FOUR HOURS AND A HALF, while that employed by the soldiers in their cooking is less than TWO HOURS AND A HALF; yet the quantity of fuel consumed by the latter is near THIRTEEN TIMES greater than that employed in the public kitchen of the House of Industry.

But I must not here anticipate here a matter which is to be the subject of a separate Essay; and which, from its great importance, certainly deserves to be carefully and thoroughly investigated.

CHAPTER. V.

All those who have been conversant in military affairs must have had frequent opportunities of observing the striking difference there is, even in the appearance of the men, between regiments in which messes are established, and Food is regularly provided under the care and inspection of the Officers; and others, in which the soldiers are left individually to shift for themselves. And the difference which may be observed between soldiers who live in messes, and are regularly fed, and others who are not, is not confined merely to their external appearance: the influence of these causes extends much farther, and even the MORAL CHARACTER of the man is affected by them.

Peace of mind, which is as essential to contentment and happiness as it is to virtue, depends much upon order and regularity in the common affairs of life; and in no case are order and method more necessary to happiness, (and consequently to virtue,) than in that, where the preservation of health is connected with the satisfying of hunger; an appetite whose cravings are sometimes as inordinate as they are insatiable.

Peace of mind depends likewise much upon economy, or the means used for preventing pecuniary embarrassments; and the savings to soldiers in providing Food, which arise from housekeeping in messes of ten or twelve persons who live together, is very great indeed.

But great as these savings now are, I think they might be made still more considerable; and I shall give my reasons for this opinion.

Though the Bavarian soldiers live at a very small expense, little more than TWO-PENCE sterling a-day, yet when I compare this sum, small as it is, with the expense of feeding the Poor in the House of Industry at Munich, which does not amount to more than TWO FARTHINGS a-day, even including the cost of the piece of dry rye-bread, weighing seven ounces Avoirdupois[11], which is given them in their hands, at dinner, but which they seldom eat at dinner, but commonly carry home in their pockets for their suppers;—when I compare, I say, this small sum, with the daily expence of the soldiers for their subsistence, I find reason to conclude, either that the soldiers might be fed cheaper, or that the Poor must be absolutely starved upon their allowance. That the latter is not the case, the healthy countenances of the Poor, and the air of placid contentment which always accompanies them, as well in the dining-hall as in their working-rooms, affords at the same time the most interesting and most satisfactory proof possible.

Were they to go home in the course of the day, it might be suspected that they got something at home to eat, in addition to what they receive from the public kitchen of the Establishment;— but this they seldom or ever do; and they come to the house so early in the morning, and leave it so late at night, that it does not seem probable that they could find time to cook any thing at their own lodgings.

Some of them, I known, make a constant practice of giving themselves a treat of a pint of beer at night, after they have finished their work; but I do not believe they have any thing else for their suppers, except it be the bread which they carry home from the House of Industry.

I must confess, however, very fairly, that it always appeared to me quite surprising, and that it is still a mystery which I do not clearly understand, how it is possible for these poor people to be so comfortably fed upon the small allowances which they receive.— The facts, however, are not only certain, but they are notorious. Many persons of the most respectable characters in this country, (Great Britain,) as well as upon the Continent, who have visited the

House of Industry at Munich, can bear witness to their authenticity; and they are surely not the less interesting for being extraordinary.

It must however be remembered, that what formerly cost TWO FARTHINGS in Bavaria, at the mean price of provisions in that country, costs THREE farthings at this present moment; and would probably cost SIX in London, and in most other parts of Great Britain: but still, it will doubtless appear almost incredible, that a comfortable and nourishing meal, sufficient for satisfying the hunger of a strong man, may be furnished in London, and at this very moment, when provisions of all kinds are so remarkably dear, at LESS THAN THREE FARTHINGS. The fact, however, is most certain, and may easily be demonstrated by making the experiment.

Supposing that it should be necessary, in feeding the Poor in this country, to furnish them with three meals a-day, even that might be done at a very small expence, were the system of feeding them adopted which is here proposed. The amount of that expence would be as follows:

<pre>
                                         Pence. Farths.
For breakfast, 20 ounces of the Soup No, II.
   composed of pearl barley, peas, potatoes,
   and fine wheaten bread (See page 210.) 0 2 1/2
For dinner, 20 ounces of the same Soup, and
   7 ounces of rye-bread … … … … 1 2
For supper, 20 ounces of the same Soup … 0 2 1/2
                                         — — — — — —
In all 4 lb. 3 oz. of Food[12], which would cost 2 3
</pre>

Should it be thought necessary to give a little meat at dinner, this may best be done by mixing it, cut fine, or minced, in bread dumplins; or when bacon, or any kind of salted or smoked meat is given, to cut it fine and mix it with the bread which is eaten in the soup. If the bread be fried, the Food will be much improved; but this will be attended with some additional expence. —Rye-bread is as good, if not better, for frying, than bread made of wheat flour; and it is commonly not half so dear.— Perhaps rye-bread fried might be furnished almost as cheap as wheaten bread not fried; and if this could be done, it would certainly be a very great improvement.

There is another way by which these cheap soups may be made exceedingly palatable and savoury;—which is by mixing with them a very small quantity of red herrings, minced very fine or pounded in a mortar.—There is no kind of cheap Food, I believe, that has so much taste as red herrings, or that communicates its flavour with so much liberality to other eatables; and to most palates it is remarkably agreeable.

Cheese may likewise be made use of for giving an agreeable relish to these soups; and a very small quantity of it will be sufficient for that purpose, provided it has a strong taste, and is properly applied.—It should be grated to a powder with a grater, and a small quantity of this powder thrown over the soup, AFTER IT IS DISHED OUT.—This is frequently done at the sumptuous tables of the rich, and is thought a great delicacy; while the Poor, who have so few enjoyments, have not been taught to avail themselves of this, which is so much within their reach.

Those whole avocations call them to visit distant countries, and those whose fortune enables them to travel for their amusement or improvement, have many opportunities of acquiring useful information; and in consequence of this intercourse with strangers, many improvements, and more REFINEMENTS, have been introduced into this country; but the most important advantages that MIGHT be derived from an intimate knowledge of the manners and customs of differing nations,—the introduction of improvements tending to facilitate the means of subsistence, and to increase the comforts and conveniences of the most necessitous and most numerous classes of society,—have been, alas! little attended to. Our extensive commerce enables us to procure, and we do actually import most of the valuable commodities which are the produce either of the foil of the ocean, or of the industry of man in all the various regions of the habitable globe;—but the result of the EXPERIENCE OF AGES respecting the use that can be made of those commodities has seldom been thought worth importing! I never see maccaroni in England, or polenta in Germany, upon the tables of the rich, without lamenting that cheap and wholesome luxuries should be monopolized by those who stand least in need of them; while the Poor, who, one would think, ought to be considered as having almost an EXCLU-

SIVE right to them, (as they were both invented by the Poor of a neighbouring nation,) are kept in perfect ignorance of them.

But these two kinds of Food are so palatable, wholesome, and nourishing, and may be provided so easily, and at so very cheap a rate in all countries, and particularly in Great Britain, that I think I cannot do better than to devote a few pages to the examination of them; — and I shall begin with Polenta, or Indian corn, as it is called in this country.

CHAPTER. VI.

Of INDIAN CORN.
It affords the cheapest and most nourishing food known.
Proofs that it is more nourishing than rice.
Different ways of preparing or cooking it.
Computation of the expense of feeding a person with it,
   founded on experiment.
Approved Receipt for making an INDIAN PUDDING.

I cannot help increasing the length of this Essay much beyond the bounds I originally assigned to it, in order to have an opportunity of recommending a kind of Food which I believe to be beyond comparison the most nourishing, cheapest, and most wholesome that can be procured for feeding the Poor. — This is Indian Corn, a most valuable production; and which grows in almost all climates; and though it does not succeed remarkably well in Great Britain, and in some parts of Germany, yet it may easily be had in great abundance, from other countries; and commonly at a very low rate.

The common people in the northern parts of Italy live almost entirely upon it; and throughout the whole Continent of America it makes a principal article of Food. — In Italy it is called Polenta, where it is prepared or cooked in a variety of ways, and forms the basis of a number of very nourishing dishes. — The most common way however of using it in that country is to grind it into meal, and with water to make it into a thick kind of pudding, like what in this country is called a hasty-pudding, which is eaten with various kinds of sauce, and sometimes without any sauce.

In the northern parts of North America, the common household bread throughout the country is composed of one part of Indian meal and one part of rye meal; and I much doubt whether a more wholesome, or more nourishing kind of bread can be made.

Rice is universally allowed to be very nourishing, — much more so even than wheat; but there is a circumstance well known to all those who are acquainted with the details of feeding the negro slaves in the southern states of North America, and in the West Indies, that would seem to prove, in a very decisive and satisfactory manner, that INDIAN CORN IS EVEN MORE NOURISHING THAN RICE. — In those countries, where rice and Indian Corn are both produced in the greatest abundance, the negroes have frequently had their option between these two kinds of Food; and have invariably preferred the latter. — The reasons they give for this preference they express in strong, though not in very delicate terms. — They say that "Rice turns to water in their bellies, and runs off;" — but "Indian Corn stays with them, and makes strong to work."

This account of the preference which negroes give to Indian Corn for Food, and of their reasons for this preference, was communicated to me by two gentlemen of most respectable character, well known in England, and now resident in London, who were formerly planters; one in Georgia, and the other in Jamaica.

The nutritive quantity which Indian Corn possessed, in a most eminent degree, when employed for fattening hogs and poultry, and for giving strength to working oxen, has long been universally known and acknowledged in every part of North America; and nobody in that country thinks of employing any other grain for those purposes.

All these facts prove to a demonstration that India Corn possesses very extraordinary nutritive powers; and it is well known that there is no species of grain that can be had so cheap, or in so great abundance; — it is therefore well worthy the attention of those who are engaged in providing cheap and wholesome Food for the Poor, — or in taking measures for warding off the evils which commonly attend a general scarcity of provisions, to consider in time, how this useful article of Food may be procured in large quantities, and how the introduction of it into common use can be most easily be effected.

In regard to the manner of using Indian Corn, there are a vast variety of different ways in which it may be prepared, or cooked, in order to its being used as Food. — One simple and obvious way of using it, is to mix it with wheat, rye, or barley meal, in making bread; but when it is used for making bread, and particularly when it is mixed with wheat flour, it will greatly improve the quality of the bread if the Indian meal, (the coarser part of the bran being first separated from it by sifting,) be previously mixed with water, and boiled for a considerable length of time,—two or three hours for instance, over a slow fire, before the other meal or flour is added to it.—This boiling, which, if the proper quantity of water is employed, will bring the mass to the consistency of a thin pudding, will effectually remove a certain disagreeable RAW TASTE in the Indian Corn, which simple baking will not entirely take away; and the wheat flour being mixed with this pudding after it has been taken from the fire and cooked, and the whole well kneaded together, may be made to rise, and be formed into loaves, and baked into bread, with the same facility that bread is made of wheat flour alone, or of any mixtures of different kinds of meal.

When the Indian meal is previously prepared by boiling, in the manner here described, a most excellent, and very palatable kind of bread, not inferior to wheaten bread, may be made of equal parts of this meal and of common wheat flour.

But the most simple, and I believe the best, and most economical way of employing Indian Corn as Food, is to make it into puddings.—There is, as I have already observed, a certain rawness in the taste of it, which nothing but long boiling can remove; but when that disagreeable taste is removed, it becomes extremely palatable; and that it is remarkably wholesome, has been proved by so much experience that no doubts can possibly be entertained of that fact.

The culture of it required more labour than most other kinds of grain; but, on the other hand, the produce is very abundant, and it is always much cheaper than either wheat or rye.— The price of it in the Carolinas, and in Georgia, has often been as low as eighteen pence, and sometimes as one shilling sterling per bushel;—but the Indian Corn which is grown in those southern states is much inferior, both in weight and in its qualities, to that which is the produce of colder climates.—Indian Corn of the growth of Canada, and the New England states, which is generally thought to be worth twenty per cent. more per bushel than that which is grown in the southern states, may commonly be bought for two and sixpence, or three shillings a bushel.

It is now three shillings and sixpence a bushel at Boston; but the prices of provisions of all kinds have been much raised of late in all parts of America, owing to the uncommonly high prices which are paid for them in the European markets since the commencement of the present war.

Indian Corn and rye are very nearly of the same weight, but the former gives rather more flour, when ground and sifted, than the latter.—I find by a report of the Board of Agriculture, of the 10th of November 1795, that three bushels of Indian Corn weighed 1 cwt. 1 qr. 18 lb. (or 53 lb. each bushel), and gave 1 cwt. 20 lb. of flour and 26 lb. of bran; while three bushels of rye, weighing 1 cwt. 1 qr. 22 lb. (or 54 lb. the bushel), gave only 1 cwt. 17 lb. of flour and 28 lb. of bran.— But I much suspect that the Indian Corn used in these experiments was not of the best quality[13].

I saw some of it, and it appeared to me to be of that kind which is commonly grown in the southern states of North America.— Indian Corn of the growth of colder climates is, probably, at least as heavy as wheat, which weights at a medium about 58 lb. per bushel, and I imagine it will give nearly as much flour[14].

In regard to the most advantageous method of using Indian Corn as Food, I would strongly recommend, particularly when it is employed for feeding the Poor, a dish made of it that is in the highest estimation throughout America, and which is really very good, and very nourishing. This is called hasty-pudding; and it is made in the following manner: A quantity of water, proportioned to the quantity of hasty-pudding intended to be made, is put over the fire in an open iron pot, or kettle, and a proper quantity of salt for seasoning the pudding being previously dissolved in the water, Indian meal is stirred into it, by little and little, with a wooded spoon with a long handle, while the water goes on to be heated and made to boil;— great care being taken to put in the meal by very small quantities, and by sifting it slowly through the fingers of the left hand, and stirring the water about very briskly at the same time with the wooden spoon, with the right hand, to mix the meal with the water in such a manner as to prevent lumps being formed.— The meal should be added so slowly, that, when the water is brought to boil, the mass should not be thicker than water-gruel, and half an hour more, at least, should be employed to add the additional quantity of meal necessary for bringing the pudding to be of the proper consistency; during which time it should be stirred about continually, and kept constantly boiling. — The method of determining when the pudding has acquired the proper consistency is this; — the wooden spoon used for stirring it being placed upright in the middle of the kettle, if it falls down, more meal must be added; but if the pudding is sufficiently thick and adhesive to support it in a vertical position, it is declared to be PROOF; and no more meal is added. — If the boiling, instead of being continued only half an hour, be prolonged to three quarters of an hour, or an hour, the pudding will be considerably improved by this prolongation.

This hasty-pudding, when done, may be eaten in various ways. — It may be put, while hot, by spoonfuls into a bowl of milk, and eaten with the milk with a spoon, in lieu of bread; and used in this way it is remarkably palatable. — It may likewise be eaten, while hot, with a sauce composed of butter and brown sugar, or butter and molasses, with or without a few drops of vinegar; and however people who have not been accustomed to this American cookery may be prejudiced against it, they will find upon trial that it makes a most excellent dish, and one which never fails to be much liked by those who are accustomed to it. — The universal fondness of Americans for it proves that it must have some merit; — for in a country which produces all the delicacies of the table in the greatest abundance, it is not to be supposed that a whole nation should have a taste so depraved as to give a decided preference to any particular species of Food which has not something to recommend it.

The manner in which hasty-pudding is eaten with butter and sugar, or butter and molasses, in America, is as follows: The hasty-pudding being spread out equally upon a plate, while hot, an excavation is made in the middle of it, with a spoon, into which excavation a piece of butter, as large as a nutmeg, is put; and upon it, a spoonful of brown sugar, or more commonly of molasses. — The butter being soon melted by the heat of the pudding, mixes with the sugar, or molasses, and forms a sauce, which, being confined in the excavation made for it, occupies the middle of the plate. — The pudding is then eaten with a spoon, each spoonful of it being dipt into the sauce before it is carried to the mouth; care being had in taking it up, to begin on the outside, or near the brim of the plate, and to approach the center by regular advances, in order not to demolish too soon the excavation which forms the reservoir for the sauce.

If I am prolix in these descriptions, my reader must excuse me; for persuaded as I am that the action of Food upon the palate, and consequently the pleasure of eating, depends very much indeed upon the MANNER in which the Food is applied to the organs of taste, I have thought it necessary to mention, and even to illustrate in the clearest manner, every circumstance which appeared to me to have influence in producing those important effects.

In the case in question, as it is the sauce alone which gives taste and palatableness to the Food, and consequently is the cause of the pleasure enjoyed in eating it, the importance of applying, or using it, in such a manner as to produce the greatest and most durable effect possible on the organs of taste, is quite evident; and in the manner of eating this Food which has here been described and recommended, the small quantity of sauce used, (and the quantity must be small, as it is the expensive article,) is certainly applied to the palate more immediately; — by a greater surface; — and in a state of greater condensation; — and consequently acts upon it more powerfully; — and continues to act upon it for a greater length of time, than it could well be made to do when used in any other way. — Were it more intimately mixed with the pudding, for instance, instead of being merely applied to its external surface, its action would certainly be much less powerful; and were it poured over the pudding, or was proper care not taken to keep it confined in the little excavation or reservoir made in the midst of the pudding to contain it, much of it would attach itself and adhere to the surface of the plate, and be lost.

Hasty-pudding has this in particular to recommend it; — and which renders it singularly useful as Food for poor families, — that when more of it is made at once than is immediately wanted, what remains may be preserved good for several days, and a number of very palatable dishes may be made of it. — It may be cut in thin slice, and toasted before the fire, or on a gridiron, and eaten instead of bread, either in milk, or in any kind of soup or pottage; or with any other kind of Food with which bread is commonly eaten; or it may be eaten cold, without any preparation, with a warm sauce made of butter, molasses, or sugar, and a little vinegar.—In this last-mentioned way of eating it, it is quite as palatable, and I believe more wholesome, than when eaten warm; that is to say, when it is first made.—It may likewise be put cold, without any preparation, into hot milk; and this mixture is by no means unpalatable, particularly if it be suffered to remain in the milk till it is warmed throughout, or if it be boiled in the milk for a few moments.

A favourite dish in America, and a very good one, is made of cold boiled cabbage chopped fine, with a small quantity of cold boiled beef, and slices of cold hasty-pudding, all fried together in butter or hog's lard.

Though hasty-puddings are commonly made of Indian meal, yet it is by no means uncommon to make them of equal parts of Indian, and of rye meal;—and they are sometimes made of rye meal alone; or of rye meal and wheat flour mixed.

To give a satisfactory idea of the expence of preparing hasty-puddings in this country, (England,) and of feeding the Poor with them, I made the following experiment:—About 2 pints of water, which weighed just 2 lb. Avoirdupois, were put over the fire in a saucepan of a proper size, and 58 grains in weight or 1/120 of a pound of salt being added, the water was made to boil.—During the time that is was heating, small quantities of Indian meal were stirred into it, and care was taken, by moving the water briskly about, with a wooden spoon, to prevent the meal from being formed into lumps; and as often as any lumps were observed, they were carefully broken with the spoon;—the boiling was then continued half an hour, and during this time the pudding was continually stirred about with the wooden spoon, and so much more meal was added as was found necessary to bring the pudding to be of the proper consistency.

This being done, it was taken from the fire and weighed, and was found to weigh just 1 lb. 11 1/2 oz.—Upon weighing the meal which remained, (the quantity first provided having been exactly determined by weight in the beginning of the experiment,) it was found that just HALF A POUND of meal had been used.

From the result of this experiment it appears, that for each pound of Indian meal employed in making hasty-pudding, we may reckon

3 lb. 9 oz. of the pudding.—And expence of providing this kind of Food, or the cost of it by the pound, at the present high price of grain in this country, may be seen by the following computation:

L. s. d.

Half a pound of Indian meal, (the quantity)]
  used in the foregoing experiment,) at 2d]
  a pound or 7s. 6d. a bushel for the corn,]... 0 0 1
  (the price stated in the report of the]
  Board of Agriculture of the 10th of]
  November 1795, so often referred to,) costs]

58 grains or 1/120 of a pound of salt, at]
2d. per pound]... 0 0 0 1/60

———————

Total, 0 0 1 1/60

Now, as the quantity of pudding prepared with these ingredients was 1 lb. 11 1/2 oz. and the cost of the ingredients amounted to ONE PENNY AND ONE SIXTIETH OF A PENNY, this gives for the cost of one pound of hasty-pudding 71/120 of a penny, or 2 1/3 farthings, very nearly.—It must however be remembered that the Indian Corn is here reckoned at a very exorbitant price indeed[15].

But before it can be determined what the expence will be of feeding the Poor with this kind of Food, it will be necessary to ascertain how much of it will be required to give a comfortable meal to one person; and how much the expence will be of providing the sauce for that quantity of pudding.—To determine these two points with some degree of precision, I made the following experiment:— Having taken my breakfast, consisting of two dishes of coffee, with cream, and a dry toast, at my usual hour of breakfasting, (nine o'clock in the morning,) and having fasted from that time till five o'clock in the afternoon, I then dined upon my hasty-pudding, with the American sauce already described, and I found, after my appetite for Food was perfectly satisfied, and I felt that I had made a comfortable dinner, that I had eaten just 1 lb. 1 1/2 oz. of the pudding; and the ingredients, of which the sauce which was eaten with it was composed, were half an ounce of butter; three quarters of an ounce of molasses; and 21 grains or 1/342 of a pint of vinegar.

The cost of this dinner may be seen by the following computation:

For the Pudding

Farthings.

1 lb. 1 1/2 oz. of hasty-pudding, at
2 1/3 farthings a pound … … … … 2 1/2

— — —

For the Sauce

Half an ounce of butter, at 10d. per pound 1 1/4
Three quarters of an ounce of molasses,
at 6d. per pound … … … … 1
1/352 of a pint of vinegar, at 2s 8d.
the gallon … … … … … … 0 1/16

— — —

Total for the Sauce, 2 5/16 farthings.

Sum total of expences for this dinner,
for the pudding and its sauce… … … 4 13/16 farthings.
Or something less than one penny farthing.

I believe it would not be easy to provide a dinner in London, at this time, when provisions of all kinds are so dear, equally grateful to the palate and satisfying to the cravings of hunger, at a smaller expence. — And that this meal was sufficient for all the purposes of nourishment appears from hence, that though I took my usual exercise, and did not sup after it, I neither felt any particular faintness, nor any unusual degree of appetite for my breakfast next morning.

I have been the more particular in my account of this experiment, to show in what manner experiments of this kind ought, in my opinion, to be conducted; — and also to induce others to engage in these most useful investigations.

It will not escape the observation of the reader, that small as the expence was of providing this dinner, yet very near one-half of that sum was laid out in purchasing the ingredients for the sauce. — But it is probable that a considerable part of that expence might be saved.—In Italy, polenta, which is nothing more than hasty-pudding made with Indian meal and water, is very frequently, and I believe commonly eaten without any sauce, and when on holidays or other extraordinary occasions they indulge themselves by adding a sauce to it, this sauce is far from expensive.—It is commonly nothing more than a very small quantity of butter spread over the flat surface of the hot polenta which is spread out thin in a large platter; with a little Parmezan or other strong cheese, reduced to a coarse powder by grating it with a grater, strewed over it.

Perhaps this Italian sauce might be more agreeable to an English palate than that commonly used in America. It would certainly be less expensive, as much less butter would be required, and as cheese in this country is plenty and cheap. But whatever may be the sauce used with Food prepared of Indian Corn, I cannot too strongly recommend the use of that grain.

While I was employed in making my experiment upon hasty-pudding, I learnt from my servant, (a Bavarian,) who assisted me, a fact which gave me great pleasure, as it served to confirm me in the opinion I have long entertained of the great merit of Indian Corn.— He assured me that polenta is much esteemed by the peasantry in Bavaria, and that it makes a very considerable article of their Food; that it comes from Italy through the Tyrol; and that it is commonly sold in Bavaria AT THE SAME PRICE AS WHEAT FLOUR! Can there be stronger proofs of its merit?

The negroes in America prefer it to rice; and the Bavarian peasants to wheat.—Why then should not the inhabitants of this island like it? It will not, I hope, be pretended, that it is in this favoured soil alone that prejudices take such deep root that they are never to be eradicated, or that there is any thing peculiar in the construction of the palate of an Englishman.

The objection that may be made to Indian Corn,—that it does not thrive well in this country,—is of no weight. The same objection might, with equal reason, be made to rice, and twenty other articles of Food now in common use.

It has ever been considered, by those versed in the science of political economy, as an object of the first importance to keep down the prices of provisions, particularly in manufacturing and commercial countries;—and if there be a country on earth where this ought to be done, it is surely Great Britain:—and there is certainly no country which has the means of doing it so much in its power.

But the progress of national improvements must be very slow, however favorable other circumstances may be, where those citizens, who, by their rank and situation in society, are destined to direct the public opinion, AFFECT to consider the national prejudices as unconquerable[16].—But to return to the subject immediately under consideration.

Though hasty-pudding is, I believe, the cheapest Food that can be prepared with Indian Corn, yet several other very cheap dishes may be made of it, which in general are considered as being more palatable, and which, most probably, would be preferred in this country; and among these, what in America is called a plain Indian pudding certainly holds the first place, and can hardly fail to be much liked by those, who will be persuaded to try it.—It is not only cheap and wholesome, but a great delicacy; and it is principally on account of these puddings that the Americans, who reside in this country, import annually for their own consumption Indian Corn from the Continent of America.

In order to be able to give the most particular and satisfactory information respecting the manner of preparing these Indian puddings, I caused one of them to be made here, (in London,) under my immediate direction, by a person born and brought up in North America, and who understands perfectly the American art of cookery in all its branches[17]. This pudding, which was allowed by competent judges who tasted it to be as good as they had ever eaten, was composed and prepared in the following manner:

Approved Receipt for making a plain Indian Pudding.

Three pounds of Indian meal (from which the bran had been separated by sifting it in a common hair sieve) were put into a large bowl, and five pints of boiling water were put to it, and the whole well stirred together; three quarters of a pound of molasses and one ounce of salt were then added to it, and these being well mixed, by stirring them with the other ingredients, the pudding was poured into a fit bag; and the bag being tied up, (an empty space being left in the bag tying it, equal to about one-sixth of its contents, for giving room for the pudding to swell,) this pudding was put into a kettle of boiling water, and was boiled six hours without intermission; the loss of the water in the kettle by evaporation during this time being frequently replaced with boiling water from another kettle.

The pudding upon being taken out of the bag weighed ten pounds and one ounce; and it was found to be perfectly done, not having the smallest remains of that raw taste so disagreeable to all palates, and particularly to those who are not used to it, which always predominates in dishes prepared of Indian meal when they are not sufficiently cooked.

As this raw taste is the only well-founded objection that can be made to this most useful grain, and is, I am persuaded, the only cause which makes it disliked by those who are not accustomed to it, I would advise those who may attempt to introduce it into common use, where it is not known, to begin with Indian (bag) puddings, such as I have here been describing; and that this is a very cheap kind of Food will be evident from the following computation:

Expense of preparing the Indian Pudding above mentioned.

Pence. Pence.
3 lb. of Indian meal at ... ... 1 1/2 ... 4 1/2
3/4 lb. of molasses at ... ... 6 ... 4 1/2
1 oz. of salt at 2d. per lb. 0 1/8

— — —

Total for the ingredients, 9 1/8

As this pudding weighed 10 1/16 lbs. and the ingredients cost nine pence and half a farthing, this gives three farthings and a half for each pound of pudding.

It will be observed, that in this computation I have reckoned the Indian meal at no more than 1 1/2d per pound, whereas in the calculation which was given to determine the expense of preparing hasty-pudding it was taken at two pence a pound. I have here reckoned it at 1 1/2d. a pound, because I am persuaded it might be had here in London for that price, and even for less. — That which has lately been imported from Boston has not cost so much; and were it not for the present universal scarcity of provisions in Europe, which has naturally raised the price of grain in North America, I have no doubt but Indian meal might be had in this country for less than one penny farthing per pound.

In composing the Indian pudding above mentioned, the molasses is charged at 6d. the pound, but that price is very exorbitant. A gallon of molasses weighing about 10 lb. commonly costs in the West Indies from 7d. to 9d. sterling; and allowing sufficiently for the expenses of freight, insurance, and a fair profit for the merchant, it certainly ought not to cost in London more than 1s. 8d. the gallon[18]; and this would bring it to 2d. per pound.

If we take the prices of Indian meal and molasses as they are here ascertained, and compute the expense of the ingredients for the pudding before mentioned, it will be as follows: —

Pence. Pence.

3 lb. of Indian meal at ... ... 1 1/4 ... 3 3/4
3/4 lb. of molasses at ... ... 2 ... 1 1/2
1 oz. of salt at 2d. per lb. 0 1/8

— — —

Total for the ingredients, 5 3/8

Now as the pudding weighed 10 1/16 lbs. this gives two farthings, very nearly, for each pound of pudding; which is certainly very cheap indeed, particularly when the excellent qualities of the Food are considered.

This pudding, which ought to come out of the bag sufficiently hard to retain its form, and even to be cut into slices, is so rich and palatable, that it may very well be eaten without any sauce; but those who can afford it commonly eat it with butter. A slice of the pudding, about half an inch, or three quarters of an inch in thickness, being laid hot upon a plate, an excavation is made in the middle of it, with the point of the knife, into which a small piece of butter, as large perhaps as a nutmeg, is put, and where it soon melts. To expedite the melting of the butter, the small piece of pudding which is cut out of the middle of the slice to form the excavation for receiving the butter, is frequently laid over the butter for a few moments, and is taken away (and eaten) as soon as the butter is melted. If the butter is not salt enough, a little salt is put into it after it is melted. The pudding is to be eaten with a knife and fork, beginning at the circumference of the slice, and approaching regularly towards the center, each piece of pudding being taken up with the fork, and dipped into the butter, or dipped into it IN PART ONLY, as is commonly the case, before it is carried to the mouth.

To those who are accustomed to view objects upon a great scale, and who are too much employed in directing what ought to be done, to descend to those humble investigations which are necessary to show HOW it is to be effected, these details will doubtless appear trifling and ridiculous; but as my mind is strongly impressed with the importance of giving the most minute and circumstantial information respecting the MANNER OF PERFORMING any operation, however simple it may be, to which people have not been accustomed, I must beg the indulgence of those who may not feel themselves particularly interested in these descriptions.

In regard to the amount of the expence for sauce for a plain Indian (bag) pudding, I have found that when butter is used for that purpose, (and no other sauce ought ever to be used with it,) half an ounce of butter will suffice for one pound of the pudding. —It is very possible to contrive matters so as to use much more;—perhaps twice, or three times as much;—but if the directions relative to the MANNER of eating this Food, which have already been given, are strictly followed, the allowance of butter here determined will be quite sufficient for the purpose for which it is designed; that is to say, for giving an agreeable relish to the pudding.—Those who are particularly fond of butter may use three quarters of an ounce of it with a pound of the pudding; but I am certain, that to use an ounce would be to waste it to no purpose whatever.

If now we reckon Irish, or other firkin butter, (which, as it is salted, is the best that can be used,) at eight pence the pound, the sauce for one pound of pudding, namely, half an ounce of butter, will cost just one farthing; and this, added to the cost of the pudding, two farthings the pound, gives three farthing for the cost by the pound of this kind of food, with its sauce; and, as this food is not only very rich and nutritive, but satisfying at the same time in a very remarkable degree, it appears how well calculated it is for feeding the Poor.

It should be remembered, that the molasses used as an ingredient in these Indian puddings, does not serve merely to give taste to them;—it acts a still more important part;—it gives what, in the language of the kitchen, is called lightness.—It is a substitute for eggs, and nothing but eggs can serve as a substitute for it, except it be treacle; which, in fact, is a kind of molasses; or perhaps coarse brown sugar, which has nearly the same properties.— It prevents the pudding from being heavy, and clammy; and without communicating to it any disagreeable sweet taste, or any thing of that flavour peculiar to molasses, gives it a richness uncommonly pleasing to the palate. And to this we may add, that it is nutritive in a very extraordinary degree.—This is a fact well known in all countries where sugar is made.

How far the laws and regulations of trade existing in this country might render it difficult to procure molasses from those places where it may be had at the cheapest rate, I know not;—nor can I tell how far the free importation of it might be detrimental to our public finances;—I cannot, however, help thinking, that it is so great an object to this country to keep down the prices of provisions, or rather to check the alarming celerity with which they are rising, that means ought to be found to facilitate the importation, and introduction into common use, of an article of Food of such extensive utility. It might serve to correct in some measure, the baleful influence of another article of foreign produce, (tea,) which is doing infinite harm in this island.

A point of great importance in preparing an Indian pudding, is to boil it PROPERLY and SUFFICIENTLY. The water must be actually boiling when the pudding is put into it; and it never must be suffered to cease boiling for a moment, till it is done; and if the pudding is not boiled full six hours, it will not be sufficiently cooked.— Its hardness, when done, will depend on the space left in the bag its expansion. The consistency of the pudding ought to be such, that it can be taken out of the bag without falling to pieces;—but it is always better, on many accounts, to make it too hard than too soft. The form of the pudding may be that of a cylinder; of rather of a truncated cone, the largest end being towards the mouth of the bag, in order that it may be got out of the bag with greater facility; or it may be made of a globular form, by tying it up in a napkin.—But whatever is the form of the pudding, the bag, or napkin in which it is to be boiled, must be wet in boiling water before the pudding, (which is quite liquid before it is boiled,) is poured into it; otherwise it will be apt to run through the cloth.

Though this pudding is so good, perfectly plain, when made according to the directions here given, that I do not thing it capable of any real improvement; yet there are various additions that may be made to it, and that frequently are made to it, which may perhaps be thought by some to render it more palatable, or otherwise to improve it. Suet may, for instance, be added, and there is no suet pudding whatever superior to it; and as no sauce is necessary with a suet pudding, the expence for the suet will be nearly balanced by the saving of butter. To a pudding of the size of that just described, in the composition of which three pounds of Indian meal were used, one pound of suet will be sufficient; and this, in general, will not cost more than from five pence to six pence, even in London;—and the butter for sauce to a plain pudding of the same size would cost nearly as much. The suet pudding will indeed be rather the cheapest of the two, for the pound of suet will add a pound in weight to the pudding;—whereas the butter will only add five ounces.

As the pudding, made plain, weighing 10 1/16 lb. cost 5 3/8 pence, the same pudding, with the addition of one pound of suet, would weigh 11 1/16 lb. and would cost 11 1/8 pence,—reckoning the suet at six pence the pound.—Hence it appears that Indian suet pudding may be made in London for about one penny a pound. Wheaten bread, which is by no means so palatable, and certainly not half so nutritive, now costs something more than three pence the pound: and to this may be added, that dry bread can hardly be eaten alone; but of suet pudding a very comfortable meal may be made without any thing else.

A pudding in great repute in all parts of North America, is what is called an apple pudding. This is an Indian pudding, sometimes with, and sometimes without suet, with dried cuttings of sweet apples mixed with it; and when eaten with butter, it is most delicious Food. These apples, which are pared as soon as they are gathered from the tree, and being cut into small pieces, are freed from their cores, and thoroughly dried in the sun, may be kept good for several years. The proportions of the ingredients used in making these apple puddings are various; but, in general, about one pound of dried apples is mixed with three pounds of meal,—three quarters of a pound of molasses,—half an ounce of salt, and five pints of boiling water.

In America, various kinds of berries, found wild in the woods, such as huckle-berries, belberries, whortle-berries, etc. are gathered and dried, and afterwards used as ingredients in Indian puddings: and dried cherries and plums may be made use of in the same manner.

All these Indian puddings have this advantage in common, that they are very good WARMED UP.—They will all keep good several days; and when cut into thin slices and toasted, are an excellent substitute for bread.

It will doubtless be remarked, that in computing the expence of providing these different kinds of puddings, I have taken no notice of the expence which will be necessary for fuel to cook them.—This is an article which ought undoubtedly to be taken into the account. The reason of my not doing it here is this:— Having, in the course of my Experiments on Heat, found means to perform all the common operations of cookery with a surprisingly small expence of fuel, I find that the expence in question, when the proper arrangements are made for saving fuel, will be very trifling. And farther, as I mean soon to publish my Treatise on the Management of Heat, in which I shall give the most ample directions relative to the mechanical arrangements of kitchen fire-places, and the best forms for all kinds of kitchen utensils, I was desirous not to anticipate a subject which will more naturally find its place in another Essay.—In the mean time I would observe, for the satisfaction of those who may have doubts respecting the smallness of the expence necessary for fuel in cooking for the Poor, that the result of many experiments, of which I shall hereafter publish a particular account, has proved in the most satisfactory manner, that when Food is prepared in large quantities, and cooked in kitchens properly arranged, the expense for fuel ought never to amount to more than two per cent. of the cost of the Food, even where victuals of the cheapest kind are provided, such as is commonly used in feeding the Poor. In the Public Kitchen of the House of Industry at Munich the expence for fuel is less than one per cent. of the cost of the Food, as may be seen in the computation, page 206, Chapter III. of this Essay: and it ought not to be greater in many parts of Great Britain.

With regard to the price at which Indian Corn can be imported into this country from North America in time of peace, the following information, which I procured through the medium of a friend, from Captain Scott, a most worthy man, who has been constantly employed above thirty years as master of a ship in the trade between London and Boston in the State of Massachusetts, will doubtless be considered as authentic[19].

The following are the questions which were put to him,—with his answers to them:

Q. What is the freight, per ton, of merchandise from Boston in North America to London in time of peace?——A. Forty shillings (sterling).

Q. What is the freight, per barrel, of Indian Corn?——A. Five shillings.

Q. How much per cent. is paid for insurance from Boston to London in time of peace?—— A. Two per cent.

Q. What is the medium price of Indian Corn, per bushel, in New England?——A. Two shillings and sixpence.

Q. What is the price of it at this time?——A. Three shillings and sixpence.

Q. How many bushels of Indian Corn are reckoned to a barrel? ——A. Four

From this account it appears that Indian Corn might, in time of peace, be imported into this country and sold here for less than four shillings the bushel;—and that it ought not to cost at this moment much more than five shillings a bushel.

If it be imported in casks, (which is certainly the best way of packing it,) as the freight of a barrel containing four bushels is five shillings, this gives 1s. 3d. a bushel for freight; and if we add one penny a bushel for insurance, this will make the amount of freight and insurance 1s. 4d. which, added to the prime cost of the Corn in America, (2s. 6d. per bushel in the time of peace, and 3s. 6d. at this time,) will bring it to 3s. 10d. per bushel in time of peace, and 4s. 10d at this present moment.

A bushel of Indian Corn of the growth of New England was found to weigh 61 lb.; but we will suppose it to weigh at a medium only 60 lb. per bushel; and we will also suppose that to each bushel of Corn when ground there is 9 lb. of bran, which is surely a very large allowance, and 1 lb. of waste in grinding and sifting;— this will leave 50 lb. of flour for each bushel of the Corn; and as it will cost, in time of peace, only 3s. 10d. or 46 pence, this gives for each pound of flour 46/50 of a penny, or 3 3/4 farthings very nearly.

If the price of the Indian Corn per bushel be taken at 4s. 10d. what it ought to cost at this time in London, without any bounty on importation being brought into the account,— the price of the flour will be 4s. 10d equal to 58 pence for 50 lb. in weight, or 1 1/6 penny the pound, which is less than one third of the present price of wheat flour. Rice, which is certainly not more nourishing than Indian Corn, costs 4 1/2 pence the pound.

If 1/13 of the value of Indian Corn be added to defray the expence of grinding it, the price of the flour will not even then be greater in London than one penny the pound in time of peace, and about one penny farthing at the present high price of that grain in North America. Hence it appears, that in stating the mean price in London of the flour of Indian Corn at one penny farthing, I have rather rated it too high than too low.

With regard to the expense of importing it, there may be, and doubtless there are frequently other expences besides those of freight and insurance; but, on the other hand, a very considerable part of the expences attending the importation of it may be reimbursed by the profits arising from the sale of the barrels in which it is imported, as I have been informed by a person who imports it every year, and always avails himself of that advantage.

One circumstance much in favour of the introduction of Indian Corn into common use in this country is the facility with which it may be had in any quantity. It grows in all quarters of the globe, and almost in every climate; and in hot countries two or three crops of it may be raised from the same ground in the course of a year.—It succeeds equally well in the cold regions of Canada;—in the temperate climes of the United States of America;—and in the burning heats of the tropics; and it might be had from Africa and Asia as well as from America. And were it even true,—what I never can be persuaded to believe,—that it would be impossible to introduce it as an article of Food in this country, it might at least be used as fodder for cattle, whose aversion to it, I will venture to say, would not be found to be UNCONQUERABLE.

Oats now cost near two pence the pound in this country. Indian Corn, which would cost but a little more than half as much, would certainly be much more nourishing, even for horses, as well as for horned cattle;—and as for hogs and poultry, they ought never to be fed with any other grain. Those who have tasted the pork and the poultry fatted on Indian Corn will readily give their assent to this opinion.

CHAPTER. VII.

When I began writing the foregoing Chapter of this Essay, I had hopes of being able to procure satisfactory information respecting the manner in which the maccaroni eaten by the Poor in Italy, and particularly in the kingdom of Naples, is prepared;— but though I have taken much pains in making these inquiries, my success in them has not been such as I could have wished:— The process, I have often been told, is very simple; and from the very low price at which maccaroni is sold, ready cooked, to the Lazzaroni in the streets of Naples, it cannot be expensive. —There is a better kind of maccaroni which is prepared and sold by the nuns in some of the convents in Italy, which is much dearer; but this sort would in any country be too expensive to be used as Food for the Poor.—It is however not dearer than many kinds of Food used by the Poor in this country; and as it is very palatable and wholesome, and may be used in a variety of ways, a receipt for preparing it may perhaps not be unacceptable to many of my readers.

A Receipt for making that Kind of Maccaroni called in Italy TAGLIATI.

Take any number of fresh-laid eggs and break them into a bowl or tray, beat them up with a spoon, but not to a froth,—add of the finest wheat flour as much as is necessary to form a dough of the consistence of paste.—Work this paste well with a rolling-pin;— roll it out into very thin leaves;—lay ten or twelve of these leaves one upon the other, and with a sharp knife cut them into very fine threads.—These threads (which, if the mass is of a proper consistency, will not adhere to each other) are to be laid on a clean board, or on paper, and dried in the air.

This maccaroni, (or cut paste as it is called in Germany, where it is in great repute,) may be eaten in various ways; but the most common way of using it is to eat it with milk instead of bread, and with chicken broth, and other broths and soups, with which it is boiled. With proper care it may be kept good for many months. It is sometimes fried in butter, and in this way of cooking it, it forms a most excellent dish indeed; inferior, I believe, to no dish of flour that can be made. It is not, however, a very cheap dish, as eggs and butter are both expensive articles in most countries.

An inferiour kind of cut paste is sometimes prepared by the Poor in Germany, which is made simply of water and wheat flour, and this has more resemblance to common maccaroni than that just described; and might, in many cases, be used instead of it. I do not think, however, that it can be kept long without spoiling; whereas maccaroni, as is well known, may be kept good for a great length of time.—Though I have not been able to get any satisfactory information relative to the process of making maccaroni, yet I have made some experiments to ascertain the expense of cooking it, and of the cost of the cheese necessary for giving it a relish.

Half a pound of maccaroni, which was purchased at an Italian shop in London, and which cost ten pence[20], was boiled till it was sufficiently done, namely, about one hour and an half, when, being taken out of the boiling water and weighed, it was found to weigh thirty-one ounces and an half, or one pound fifteen ounces and an half. The quantity of cheese employed to give a relish to this dish of boiled maccaroni, (and which was grated over it after it was put into the dish,) was one ounce, and cost two farthings.

Maccaroni is considered as very cheap Food in those countries where it is prepared in the greatest perfection, and where it is in common use among the lower classes of society; and as wheat, of which grain it is always made, is a staple commodity in this country, it would certainly be worth while to take some trouble to introduce the manufacture of it, particularly as it is already become an article of luxury upon the tables of the rich, and as great quantities of it are annually imported and sold here at a most exorbitant price[21]:—But maccaroni is by no means the cheapest Food that can be provided for feeding the Poor, in this island;—nor do I believe it is so in any country.—Polenta, or Indian Corn, of which so much has already been said,— and Potatoes, of which too much cannot be said,—are both much better adapted, in all respects, for that purpose.—Maccaroni would however, I am persuaded, could it be prepared in this country, be much less expensive than many kinds of Food now commonly used by our Poor; and consequently might be of considerable use to them.

With regard to Potatoes they are now so generally known and their usefulness is so universally acknowledged, that it would be a waste of time to attempt to recommend them.—I shall therefore content myself with merely giving receipts for a few cheap dishes in which they are employed as a principal ingredient.

Though there is no article used as Food of which a greater variety of well-tasted and wholesome dishes may be prepared than of potatoes, yet it seems to be the unanimous opinion of those who are most acquainted with these useful vegetables, that the best way of cooking them is to boil them simply, and with their skins on, in water.—But the manner of boiling them is by no means a matter of indifference.—This process is better understood in Ireland, where by much the greater part of the inhabitants live almost entirely on this Food, than any where else.

This is what might have been expected;—but those who have never considered with attention the extreme slowness of the progress of national improvements, WHERE NOBODY TAKES PAINS TO ACCELERATE THEM, will doubtless be surprised when they are told that in most parts of England, though the use of potatoes all over the country has for so many years been general, yet, to this hour, few, comparatively, who eat them, know how to dress them properly.— The inhabitants of those countries which lie on the sea-coast opposite to Ireland have adopted the Irish method of boiling potatoes; but it is more than probable that a century at least would have been required for those improvements to have made their way through the island, had not the present alarms on account of a scarcity of grain roused the public, and fixed their attention upon a subject too long neglected in this enlightened country.

The introduction of improvements tending to increase the comforts and innocent enjoyments of that numerous and useful class of mankind who earn their bread by the sweat of their brow, is an object not more interesting to a benevolent mind than it is important in the eyes of an enlightened statesman.

There are, without doubt, GREAT MEN who will smile at seeing these observations connected with a subject so humble and obscure as the boiling of potatoes, but GOOD MEN will feel that the subject is not unworthy of their attention.

The following directions for boiling potatoes, which I have copied from a late Report of the Board of Agriculture, I can recommend from my own experience:

On the boiling of Potatoes so as to be eat as Bread.

There is nothing that would tend more to promote the consumption of potatoes than to have the proper mode of preparing them as Food generally known.—In London, this is little attended to; whereas in Lancashire and Ireland the boiling of potatoes is brought to very great perfection indeed. When prepared in the following manner, if the quality of the root is good, they may be eat as bread, a practice not unusual in Ireland.—The potatoes should be, as much as possible, of the same size, and the large and small ones boiled separately.—They must be washed clean, and, without paring or scraping, put in a pot with cold water, not sufficient to cover them, as they will produce themselves, before they boil, a considerable quantity of fluid.—They do not admit being put into a vessel of boiling water like greens. — If the potatoes are tolerably large, it will be necessary, as soon as they begin to boil, to throw in some cold water, and occasionally to repeat it, till the potatoes are boiled to the heart, (which will take from half an hour to an hour and a quarter, according to their size,) they will otherwise crack, and burst to pieces on the outside, whilst the inside will be nearly in a crude state, and consequently very unpalatable and unwholesome. — During the boiling, throwing in a little salt occasionally is found a great improvement, and it is certain that the slower they are cooked the better. — When boiled, pour off the water, and evaporate the moisture, by replacing the vessel in which the potatoes were boiled once more over the fire. — This makes them remarkably dry and mealy. — They should be brought to the table with the skins on, and eat with a little salt, as bread. — Nothing but experience can satisfy any one how superior the potatoe is, thus prepared, if the sort is good and mealy. — Some prefer roasting potatoes; but the mode above detailed, extracted partly from the interesting paper of Samuel Hayes, Esquire, of Avondale, in Ireland, (Report on the Culture of Potatoes, P. 103.), and partly from the Lancashire reprinted Report (p.63.), and other communications to the Board, is at least equal, if not superior. — Some have tried boiling potatoes in steam, thinking by that process that they must imbibe less water. — But immersion in water causes the discharge of a certain substance, which the steam alone is incapable of doing, and by retaining which, the flavour of the root is injured, and they afterwards become dry by being put over the fire a second time without water. — With a little butter, or milk, of fish, they make an excellent mess.

These directions are so clear, that it is hardly possible to mistake them; and those who follow them exactly will find their potatoes surprisingly improved, and will be convinced that the manner of boiling them is a matter of much greater importance than has hitherto been imagined.

Were this method of boiling potatoes generally known in countries where these vegetables are only beginning to make their way into common use, — as in Bavaria, for instance, — I have no doubt but it would contribute more than any thing else to their speedy introduction.

The following account of an experiment, lately made in one of the parishes of this metropolis (London), was communicated to me by a friend, who has permitted me to publish it. — It will serve to show, — what I am most anxious to make appear, — that the prejudices of the Poor in regard to their Food ARE NOT UNCONQUERABLE February 25th, 1796.

The parish officers of Saint Olaves, Southwark, desirous of contributing their aid towards lessening the consumption of wheat, resolved on the following succedaneum for their customary suet puddings, which they give to their Poor for dinner one day in the week; which was ordered as follows:

<pre>
                              L. s. d.
200 lb. potatoes boiled, and
    skinned and mashed 0 8 0
2 gallons of milk 0 2 4
12 lb. of suet, at 4 1/2 ... 0 4 6
1 peck of flour 0 4 0
Baking 0 1 8
                          — — — — —-
           Expense 1 0 6
                          — — — — —-
</pre>

Their ordinary suet pudding had been made thus:

<pre>
2 bushels of flour 1 12 0
12 lb. suet 0 4 6
Baking 0 1 8
                          — — — — —-
           Expense 1 18 2
Cost of the ingredients for the
potatoes suet pudding 1 0 6
                          — — — — —-
           Difference 0 17 8
                          — — — — —-
</pre>

This was the dinner provided for 200 persons, who gave a decided perference to the cheapest of these preparations, and with it to be continued.

The following baked potatoe-puddings were prepared in the hotel where I lodge, and were tasted by a number of persons, who found them in general very palatable.

Baked Potatoe-puddings.

No. I.

12 ounces of potatoes, boiled, skinned, and mashed;
1 ounce of suet;
1 ounce (or 1/16 of a pint) of milk, and
1 ounce of Gloucester cheese.
—
Total 15 ounces, — mixed with as much boiling water as was necessary
to bring it to a due consistence, and then baked in an earthen pan.

No. II.

12 ounces of mashed potatoes as before;
1 ounces of milk, and
1 ounce of suet, with a sufficient quantity of salt. — Mixed up
with boiling water, and baked in a pan.

No. III.

12 ounces of mashed potatoes;
1 ounce of suet;
1 ounce of red herrings pounded fine in a mortar. — Mixed — baked,
etc. as before.

No. IV.

12 ounces of mashed potatoes;
1 ounce of suet, and
1 ounce of hung beef grated fine with a grater. — Mixed and baked
as before.

These puddings when baked weighed from 11 to 12 ounces each. — They were all liked by those who tasted them, but No I and No 3 seemed to meet with the most general approbation.

Receipt for a very cheap Potatoe-dumplin.

Take any quantity of potatoes, half boiled; — skin or pare them, and grate them to a coarse powder with a grater; — mix them up with a very small quantity of flour, 1/16, for instance, of the weight of the potatoes, or even less; — add a seasoning of salt, pepper, and sweet herbs; — mix up the whole with boiling water to a proper consistency, and form the mass into dumplins of the size of a large apple. — Roll the dumplins, when formed, in flour, to prevent the water from penetrating them, and put them into boiling water, and boil them till they rise to the surface of the water, and swim, when they will be found to be sufficiently done.

These dumplins may be made very savoury by mixing with them a small quantity of grated hung beef, or of pounded red herring.

Fried bread may likewise be mixed with them, and this without any other addition, except a seasoning of salt, forms an excellent dish.

Upon the same principles upon which these dumplins are prepared large boiled bag-puddings may be made; and for feeding the Poor in a public establishment, where great numbers are to be fed, puddings, as these is less trouble in preparing them, are always to be preferred to dumplins.

It would swell this Essay, (which has already exceeded the limits assigned to it,) to the size of a large volume, were I to give receipts for all the good dishes that may be prepared with potatoes. — There is however one method of preparing potatoes much in use in many parts of Germany, which appears to me to deserve being particularly mentioned and recommended; — it is as follows:

A Receipt for preparing boiled Potatoes with a Sauce.

The potatoes being properly boiled, and skinned, are cut into slices, and put into a dish, and a sauce, similar to that commonly used with a fricaseed chicken, is poured over them.

This makes an excellent and a very wholesome dish, but more calculated, it is true, for the tables of the opulent than for the Poor.—Good sauces might however be composed for this dish which would not be expensive.—Common milk-porridge, made rather thicker than usual, with wheat flour, and well salted, would not be a bad sauce for it.

Potatoe Salad.

A dish in high repute in some parts of Germany, and which deserves to be particularly recommended, is a salad of potatoes. The potatoes being properly boiled and skinned, are cut into thin slices, and the same sauce which is commonly used for salads of lettuce is poured over them; some mix anchovies with this sauce, which gives it a very agreeable relish, and with potatoes it is remarkably palatable.

Boiled potatoes cut in slices and fried in butter, or in lard, and seasoned with salt and pepper, is likewise a very palatable and wholesome dish.

Of Barley.

I have more than once mentioned the extraordinary nutritive powers of this grain, and the use of it in feeding the Poor cannot be too strongly recommended.—It is now beginning to be much used in this country, mixed with wheat flour, for making bread; but is not, I am persuaded, in bread, but in soups, that Barley can be employed to the greatest advantage.—It is astonishing how much water a small quantity of Barley-meal will thicken, and change to the consistency of a jelly; and, if my suspicions with regard to the part which water acts in nutrition are founded, this will enable us to account, not only for the nutritive quality of Barley, but also for the same quality in a still higher degree which sago and salope are known to possess.— Sago and Salope thicken, and change to the consistency of a jelly, (and as I suppose, prepare for decomposition,) a greater quantity of water than Barley, and both sago and salope are known to be nutritious in a very extraordinary degree.

Barley will thicken and change to a jelly much more water than any other grain with which we are acquainted, rice even not excepted;—and I have found reason to conclude from the result of innumerable experiments, which in the course of several years have been made under my direction in the public kitchen of the House of Industry at Munich, that for making soups, Barley is by far the best grain that can be employed.

Were I called upon to give an opinion in regard to the comparative nutritiousness of Barley-meal and wheat flour, WHEN USED IN SOUPS I should not hesitate to say that I think the former at least three or four times as nutritious as the latter.

Scotch broth is known to be one of the most nourishing dishes in common use; and there is no doubt but it owes its extraordinary nutritive quality to the Scotch (or Pearl) Barley, which is always used in preparing it.—If the Barley be omitted, the broth will be found to be poor and washy, and will afford little nourishment;—but any of the other ingredients may be retrenched;— even the meat;— without impairing very sensibly the nutritive quality of the Food.—Its flavour and palatableness may be impaired by such retrenchments; but if the water be well thickened with the Barley, the Food will still be very nourishing.

In preparing the soup used in feeding the Poor in the House of Industry at Munich, Pearl Barley has hitherto been used; but I have found, by some experiments I have lately made in London, that Pearl Barley is by no means necessary, as common Barley-meal will answer, to all intents and purposes, just as well.—In one respect it answers better, for it does not require half so much boiling.

In comparing cheap soups for feeding the Poor, the following short and plain directions will be found to be useful:

General Directions for preparing cheap Soup.

First, Each portion of Soup should consist of one pint and a quarter, which, if the Soup be rich, will afford a good meal to a grown person.—Such a portion will in general weigh about one pound and a quarter, or twenty ounces Avoirdupois.

Secondly, The basis of each portion of Soup should consist of one ounce and a quarter of Barley-meal, boiled with ONE PINT AND A QUARTER OF WATER till the whole be reduced to the uniform consistency of a thick jelly.—All other additions to the Soup do little else than to serve to make it more palatable; or by rendering a long mastication necessary, to increase and prolong the pleasure of eating; — both these objects are however of very great importance, and too much attention cannot be paid to them; but both of them may, with proper management, be attained without much expence.

Were I asked to give a Receipt for the cheapest Food which (in my opinion) it would be possible to provide in this country, it would be the following:

Receipt for a very cheap Soup.

Take of water eight gallons, and mixing with it 5 lb. of Barley-meal, boil it to the consistency of a thick jelly. — Season it with salt, pepper, vinegar, sweet herbs, and four red herrings, pounded in a mortar. — Instead of bread, add to it 5 lb. of Indian Corn made into Samp, and stirring it together with a ladle, serve it up immediately in portions of 20 ounces.

Samp, which is here recommended, is a dish said to have been invented by the savages of North America, who have no Corn-mills. — It is Indian Corn deprived of its external coat by soaking it ten or twelve hours in a lixivium of water and wood-ashes. — This coat, or husk, being separated from the kernel, rises to the surface of the water, while the grain, which is specifically heavier than water, remains at the bottom of the vessel; which grain, thus deprived of its hard coat of armour, is boiled, or rather simmered for a great length of time, two days for instance, in a kettle of water placed near the fire. — When sufficiently cooked, the kernels will be found to be swelled to a great size and burst open, and this Food, which is uncommonly sweet and nourishing, may be used in a great variety of ways; but the best way of using it is to mix it with milk, and with soups, and broths, as a substitute for bread. It is even better than bread for these purposes, for besides being quite as palatable as the very best bread, as it is less liable than bread to grow too soft when mixed with these liquids, without being disagreeably hard, it requires more mastication, and consequently tends more to increase and prolong the pleasure of eating.

The Soup which may be prepared with the quantities of ingredients mentioned in the foregoing Receipt will be sufficient for 64 portions, and the cost of these ingredients will be as follows:

Pence.

For 5 lb. of Barley-meal, at 1 1/2 pence, the]
Barley being reckoned at the present]
very high price of it in this country, viz]... 7 1/2
5s. 6d. per bushel]
5 lb. of Indian Corn, at 1 1/4 pence the pound ... 6 1/4
4 red herrings 3
Vinegar... 1
Salt 1
Pepper and sweet herbs 2

— — —-

Total 20 3/4

— — —-

This sum, (20 3/4 pence,) divided by 64, the number of portions of Soup, gives something less than ONE THIRD OF A PENNY for the cost of each portion.—But at the medium price of Barley in Great Britain, and of Indian Corn as it may be afforded here, I am persuaded that this Soup may be provided at one farthing the portion of 20 ounces.

There is another kind of Soup in great repute among the poor people, and indeed among the opulent farmers, in Germany, which would not come much higher.—This is what is called burnt Soup, or as I should rather call it, brown Soup, and it is prepared in the following manner:

Receipt for making BROWN SOUP.

Take a small piece of butter and put it over the fire in a clean frying-pan made of iron (not copper, for that metal used for this purpose would be poisonous);— put to it a few spoonfuls of wheat or rye meal;—stir the whole about briskly with a broad wooden spoon, or rather knife, with a broad and thin edge, till the butter has disappeared, and the meal is uniformly of a deep brown colour; great care being taken, by stirring it continually, to prevent the meal from being burned to the pan.

A very small quantity of this roasted meal, (perhaps half an ounce in weight would be sufficient,) being put into a sauce-pan and boiled with a pint and a quarter of water, forms a portion of Soup, which, when seasoned with salt, pepper, and vinegar, and eaten with bread cut fine, and mixed with it at the moment when it is served up, makes a kind of Food by no means unpalatable; and which is said to be very wholesome.

As this Soup may be prepared in a very short time, an instant being sufficient for boiling it; and as the ingredients for making it are very cheap, and may be easily transported, this Food is much used in Bavaria by our wood-cutters, who go into the mountains far from any habitations to fell wood.— Their provisions for a week, (the time they commonly remain in the mountains,) consist of a large loaf of rye bread (which, as it does not so soon grow dry and stale as wheaten bread, is always preferred to it); a linen bag containing a small quantity of roasted meal;—another small bag of salt;—and a small wooden box containing some pounded black pepper;—with a small frying-pan of hammered iron, about ten or eleven inches in diameter, which serves them both as an utensil for cooking, and as a dish for containing the victuals when cooked.—They sometimes, but not often, take with them a small bottle of vinegar;—but black-pepper is an ingredient in brown Soup which is never omitted.— Two table-spoonfuls of roasted meal is quite enough to make a good portion of Soup for one person; and the quantity of butter necessary to be used in roasting this quantity of meal is very small, and will cost very little.—One ounce of butter would be sufficient for roasting eight ounces of meal; and if half an ounce of roasted meal is sufficient for making one portion of Soup, the butter will not amount to more than 1/10 of an ounce; and, at eight pence the pound, will cost only 1/32 of a penny, or 1/8 of a farthing.—The cost of the meal for a portion of this Soup is not much more considerable. If it be rye meal, (which is said to be quite as good for roasting as the finest wheat flour,) it will not cost, in this country, even now when grain is so dear, more than 1 1/2d. per pound;— 1/2 an ounce, therefore, the quantity required for one portion of the Soup, would cost only 6/32 of a farthing;—and the meal and butter together no more than (1/8 + 6/32) = 10/32, or something less than 1/3 of a farthing.—If to this sum we add the cost of the ingredients used to season the Soup, namely, for salt, pepper and vinegar, allowing for them as much as the amount of the cost of the butter and the meal, or 1/3 of a farthing, this will give 2/3 of a farthing for the cost of the ingredients used in preparing one portion of this Soup; but as the bread which is eaten with it is an expensive article, this Food will not, upon the whole, be cheaper than the Soup just mentioned; and it is certainly neither so nourishing nor so wholesome.

Brown Soup might, however, on certain occasions, be found to be useful. As it is so soon cooked, and as the ingredients for making it are so easily prepared, preserved, and transported from place to place; it might be useful to travellers, and to soldiers on a march. And though it can hardly be supposed to be of itself very nourishing, yet it is possible it may render the bread eaten with it not only more nutritive, but also more wholesome;— and it certainly renders it more savoury and palatable.—It is the common breakfast of the peasants in Bavaria; and it is infinitely preferable, in all respects, to that most pernicious wash, TEA, with which the lower classes of the inhabitants of this island drench their stomachs, and ruin their constitutions.

When tea is mixed with a sufficient quantity of sugar and good cream;—when it is taken with a large quantity of bread and butter, or with toast and boiled eggs;—and above all,—WHEN IT IS NOT DRANK TOO HOT, it is certainly less unwholesome; but a simple infusion of this drug, drank boiling hot, as the Poor usually take it, is certainly a poison which, though it is sometimes slow in its operation, never fails to produce very fatal effects, even in the strongest constitution, where the free use of it is continued for a considerable length of time.

Of Rye Bread

The prejudice in this island against bread made of Rye, is the more extraordinary, as in many parts of the country no other kind of bread is used; and as the general use of it in many parts of Europe, for ages, has proved it to be perfectly wholesome.— In those countries where it is in common use, many persons prefer it to bread made of the best wheat flour; and though wheaten bread is commonly preferred to it, yet I am persuaded that the general dislike of it, where it is not much in use, is more owing to its being BADLY PREPARED, or not well baked, than to any thing else.

As an account of some experiments upon baking Rye Bread, which were made under my immediate care and inspection in the bake-house of the House of Industry at Munich, may perhaps be of use to those who wish to known how good Rye Bread may be prepared; as also to such as are desirous of ascertaining, by similar experiments, what, in any given case, the profits of a baker really are; I shall publish an account in detail of these experiments, in the Appendix to this volume.

I cannot conclude this Essay, without once more recommending, in the most earnest manner, to the attention of the Public, and more especially to the attention of all those who are engaged in public affairs,—the subject which has here been attempted to be investigated. It is certainly of very great importance, in whatever light it is considered; and it is particularly so at the present moment: for however statesmen may differ in opinion with respect to the danger or expediency of making any alterations in the constitution, or established forms of government, in times of popular commotion, no doubts can be entertained with respect to the policy of diminishing, as much as possible, at all times, —and more especially in times like the present,—the misery of the lower classes of the people.

END OF THE THIRD ESSAY.

Footnotes for Essay III.

[1] November 1795.

[2] The preparation of water is, in many cases, an object of more importance than is generally imagined; particularly when it is made use of as a vehicle for conveying agreeable tastes. In making punch, for instance, if the water used be previously boiled two or three hours with a handful of rice, the punch made from it will be incomparably better, than is to say, more full and luscious upon the palate, than when the water is not prepared.

[3] I cannot dismiss this subject, the feeding of cattle, without just mentioning another practice common among our best farmers in Bavaria, which, I think, deserves to be known. They chop the green clover with which they feed their cattle, and mix with it a considerable quantity of chopped straw. They pretend that this rich succulent grass is of so clammy a nature, that unless it be mixed with chopped straw, hay, or some other dry fodder, cattle which are fed

with it do not ruminate sufficiently. The usual proportion of the clover to the straw, is as two to one.

[4] A viertl is the twelfth part of a schafl, and the Bavarian schafl is equal to 6 31/300 Winchester bushels.

[5] The quantity of fuel here mentioned, though it certainly is almost incredibly small, was nevertheless determined from the results of actual experiments. A particular account of these experiments will be given in my Essay on the Management of Heat and the Economy of Fuel.

[6] One Bavarian schafl (equal to 6 31/100 Winchester bushels) of barley, weighing at a medium 250 Bavarian pounds, upon being pearled, or rolled (as it is called in Germany), is reduced to half a schafl, which weighs 171 Bavarian pounds. The 79lb. which it loses in the operation is the perquisite of the miller, and is all he receives for his trouble.

[7] Since the First Edition of this Essay was published the experiment with barley-meal has been tried, and the meal has been found to answer quite as well as pearl barley, if not better, for making these soups. Among others, Thomas Bernard, Esq. Treasurer of the Founding Hospital, a gentleman of most respectable character, and well known for his philanthropy and active zeal in relieving the distresses of the Poor, has given it a very complete and fair trial; and he found, what is very remarkable, though not difficult to be accounted for—that the barley-meal, WITH ALL THE BRAN IN IT, answered better, that is to say, made the soup richer, and thicker, than when the fine flour of barley, without the bran, was used.

[8] By some experiments lately made it has been found that the soup will be much improved if a small fire is made under the boiler, just sufficient to make its contents boil up once, when the barley and water are put into it, and then closing up immediately the ash-hole register, and the damper in the chimney, and throwing a thick blanket, or a warm covering over the cover of the boiler, the whole be kept hot till the next morning. This heat so long continued, acts very powerfully on the barley, and causes it to thicken the water in a very surprising manner. Perhaps the oat-meal used for making water gruel might be improved in its effects by the same means. The experiment is certainly worth trying.

[9] This invention of double bottoms might be used with great success by distillers, to prevent their liquor, when it is thick, from burning to the bottoms of their stills. But there is another hint, which I have long wished to give distillers, from which, I am persuaded, they might derive very essential advantages.—It is to recommend to them to make up warm clothing of thick blanketing for covering up their still-heads, and defending them from the cold air of the atmosphere; and for covering in the same manner all that part of the copper or boiler which rises above the brick-work in which it is fixed. The great quantity of heat is constantly given off to the cold air of the atmosphere in contact with it by this naked copper, not only occasions a very great loss of heat, and of fuel, but tends likewise very much to EMBARRASS and to PROLONG the process of distillation; for all the heat communicated by the naked still-head to the atmosphere is taken from the spirituous vapour which rises from the liquor in the still; and as this vapour cannot fail to be condensed into spirits whenever and WHEREVER it loses ANY PART of its heat,— as the spirits generated in the still-head in consequence of this communication of heat to the atmosphere do not find their way into the worm, but trickle down and mix again with the liquor in the still,—the bad effects of leaving the still-head exposed naked to the cold air is quite evident. The remedy for this evil is as cheap and as effectual, as it is simple and obvious.

[10] The Bavarian pound (equal to 1.238, or near one pound and a quarter Avoirdupois,) is divided into 32 loths.

[11] For each 100 lb. Bavarian weight, (equal to 123.84 lb. Avoirdupois,) of rye-meal, which the baker receives from the magazine, he is obliged to deliver sixty-four loaves of bread, each loaf weighing 2 lb. 5 1/2 loths; equal to 2 lb. 10 oz. Avoirdupois;—and as each loaf is divided into six portions, this gives seven ounces Avoirdupois for each portion. Hence it appears that 100 lb. of rye-meal give 149 lb. of bread; for sixty-four loaves, at 2 lb. 5 1/2 loths each, weigh 149 lb. —When this bread is reckoned at two creutzers a Bavarian pound, (which is about what it costs at a medium,) one portion costs just 10/16 of a creutzer, or 120/528 of a penny sterling, which is something less than one farthing.

[12] This allowance is evidently much too large; but I was willing to show what the expence of feed the Poor would be at THE HIGH-EST CALCULATION. I have estimated the 7 ounces of rye-bread, mentioned above, at what it ought to cost when rye is 7s. 6d. the bushel, its present price in London.

[13] Farther inquiries which have since been made, have proved that these suspicions were not without foundation.

[14] Since writing the above, I have had an opportunity of ascertaining, in the most decisive and satisfactory manner, the facts relative to the weight of Indian Corn of the growth of the northern states of America. A friend of mine, an American gentleman, resident in London, (George Erving, Esq. of Great George street, Hanover-square,) who, in common with the rest of his countrymen, still retains a liking for Indian Corn, and imports it regularly every year from America, has just received a fresh supply of it, by one of the last ships which has arrived from Boston in New England; and at my desire he weighed a bushel of it, and found it to weigh 61 lb.: It cost him at Boston three shillings and sixpence sterling the bushel.

[15] The price of Indian meal as it here estimated, — (2d. a pound,) — is at least twice as much as it would cost in Great Britain in common years, if care was taken to import it at the cheapest rate.

[16] Those who dislike trouble, and feel themselves called upon by duty and honor to take an active part in undertakings for the public good, are extremely apt to endeavour to excuse, — to themselves as well as to the world, — their inactivity and supineness, by representing the undertaking in question as being so very difficult as to make all hope of success quite chimerical and ridiculous.

[17]
The Housekeeper of my friend and countryman, Sir William Pepperel,
Bart. of Upper Seymour Street, Portman Square.

[18]
Molasses imported from the French West India Islands into the American States is commonly sold there from 12d. to 14d. the gallon.

[19] This gentleman, who is as remarkable for his good fortune at sea, as he is respectable on account of his private character and professional knowledge, has crossed the Atlantic Ocean the almost incredible number of ONE HUNDRED AND TEN TIMES! and without meeting with the smallest accident. He is now on the seas in his way to North America; and this voyage, which is his HUNDRED AND ELEVENTH, he intends should be his last. May he arrive safe,—and may he long enjoy in peace and quite the well-earned fruits of his laborious life! Who can reflect on the innumerable storms he must have experienced, and perils he has escaped, without feeling much interested in his preservation and happiness?

[20] This maccaroni would not probably have cost one quarter of that sum at Naples.—Common maccaroni is frequently sold there as low as fourteen grains, equal to five pence halfpenny sterling the rottolo, weighing twenty-eight ounces and three quarters Avoirdupois, which is three pence sterling the pound Avoirdupois. An inferiour kind of maccaroni, such as is commonly sold at Naples to the Poor, costs not more than two pence sterling the pound Avoirdupois.

[21] If maccaroni could be made in this country as cheap as it is made in Naples, that is to say, so as to be afforded for three pence sterling the pound Avoirdupois, for the best sort, (and I do not see why it should not,) as half a pound of dry maccaroni weighs when boiled very nearly two pounds, each pound of boiled maccaroni would cost only three farthings, and the cheese necessary for giving it a relish one farthing more, making together one penny; which is certainly a very moderate price for such good and wholesome Food.

CONTENTS of ESSAY IV.

of CHIMNEY FIRE-PLACES, with PROPOSALS for improving them to save FUEL; to render dwelling-houses more COMFORTABLE and SALUBRIOUS, and effectually to prevent CHIMNIES from SMOKING.

ADVERTISEMENT

CHAPTER. I. Fire-places for burning coals, or wood, in an open chimney, are capable of great improvement. Smoking chimnies may in all cases be completely cured. The immoderate size of the throats of chimnies the principal cause of all their imperfections. Philosophical investigation of the subject. Remedies proposed for all the defects that have been discovered in chimnies and their open fire-places. These remedies applicable to chimnies destined for burning wood, or turf, as well as those constructed for burning coals.

CHAPTER. II. Practical directions designed for the use of workmen, showing how they are to proceed in making the alterations necessary to improve chimney fire-places, and effectually to cure smoking chimnies.

CHAPTER. III. Of the cause of the ascent of smoke. Illustration of the subject by familiar comparisons and experiments. Of chimnies which affect and cause each other to smoke. Of chimnies which smoke from want of air. Of the eddies of wind which sometimes blow down chimnies, and cause them to smoke. Explanation of the figures.

ESSAY IV.

ADVERTISEMENT

The Author thinks it his duty to explain the reasons which have induced him to change the order in which the publication of his Essays has been announced to the Public. — Being suddenly called upon to send to Edinburgh a person acquainted with the method of altering Chimney Fire-places, which has lately been carried into execution in a number of houses in London, in order to introduce these improvements in Scotland, he did not think it prudent to send any person on so important an errand without more ample instruction than could well be given verbally; and being obliged to write on the subject, he thought it best to investigate the matter thoroughly, and to publish such particular directions respecting the improvements in question as may be sufficient to enable all those, who may be desirous of adopting them, to make, or direct the necessary alterations in their Fire-places without any further assistance.

The following Letter, which the Author received from Sir John Sinclair, Baronet, Member of Parliament, and President of the Board of Agriculture, will explain this matter more fully:

You will hear with pleasure that your mode of altering Chimnies, so as to prevent their smoking, to save fuel, and to augment heat, has answered not only with me, but with many of my friends who have tried it; and that the Lord Provest and Magistrates of Edinburgh have voted a sum of money to defray the expences of a bricklayer, who is to be sent there for the purpose of establishing the same plan in that city. I hope that you will have the goodness to expedite your paper upon the management of Heat, that the knowledge of so useful an art may be as rapidly and as extensively diffused as possible. — With my best wishes for your success in the various important pursuits in which you are now engaged, believe me, with great truth and regard, Your faithful and obedient servant John Sinclair Whitehall, London, 9th February 1796.

CHAPTER. I.

Fire-places for burning coals, or wood, in an open chimney,
  are capable of great improvement.
Smoking chimnies may in all cases be completely cured.
The immoderate size of the throats of chimnies the principal
  cause of all their imperfections.
Philosophical investigation of the subject.
Remedies proposed for all the defects that have been discovered
  in chimnies and their open fire-places.
These remedies applicable to chimnies destined for burning
  wood, or turf, as well as those constructed for burning coals.

The plague of a smoking Chimney is proverbial; but there are
many other very great defects in open Fire-places, as they are now
commonly constructed in this country, and indeed throughout Europe, which, being less obvious, are seldom attended to; and there
are some of them very fatal in their consequences to health; and, I
am persuaded, cost the lives of thousands every year in this island.

Those cold and chilling draughts of air on one side of the body,
while the other side is scorched by a Chimney Fire, which every one
who reads this must often have felt, cannot but be highly detrimental to health; and in weak and delicate constitutions must often
produce the most fatal effects. — I have not a doubt in my own mind
that thousands die in this country every year of consumptions occasioned solely by this cause. — By a cause which might be so easily
removed! — by a cause whose removal would tend to promote comfort and convenience in so many ways.

Strongly impressed as my mind is with the importance of this
subject, it is not possible for me to remain silent. — The subject is too
nearly connected with many of the most essential enjoyments of life
not to be highly interesting to all those who feel pleasure in promoting, or in contemplating the comfort and happiness of mankind. —
And without suffering myself to be deterred, either by the fear of
being thought to give the subject a degree of importance to which it
is not entitled, or by the apprehension of being tiresome to my read-

ers by the prolixity of my descriptions,—I shall proceed to investigate the subject in all its parts and details with the utmost care and attention. —And first with regard to smoking Chimnies:

There are various causes by which Chimnies may be prevented from carrying smoke; but there are none that may not easily be discovered and completely removed.—This will doubtless be considered as a bold assertion; but I trust I shall be able to make it appear in a manner perfectly satisfactory to my readers that I have not ventured to give this opinion but upon good and sufficient grounds.

Those who will take the trouble to consider the nature and properties of elastic fluids,—of air,—smoke,—and vapour,— and to examine the laws of their motions, and the necessary consequences of their being rarified by heat, will perceive that it would be as much a miracle if smoke should not rise in a Chimney, (all hindrances to its ascent being removed,) as that water should refuse to run in a syphon, or to descend in a river.

The whole mystery, therefore, of curing smoking Chimnies is comprised in this simple direction. —FIND OUT AND REMOVE THOSE LOCAL HINDRANCES WHICH FORCIBLY PREVENT THE SMOKE FROM FOLLOWING ITS NATURAL TENDENCY TO GO UP THE CHIMNEY; or rather, to speak more accurately, which prevents its being forced up the Chimney by the pressure of the heavier air of the room.

Although the causes, by which the ascent of smoke in a Chimney MAY BE obstructed, are various, yet that cause which will most commonly, and I may say almost universally be found to operate, is one which it is always very easy to discover, and as easy to remove,—the bad construction of the Chimney IN THE NEIGHBOURHOOD OF THE FIRE-PLACE.

In the course all my experience and practice in curing smoking Chimnies,—and I certainly have not had less than five hundred under my hands, and among them many which were thought to be quite incurable,—I have never been obliged, except in one single instance, to have recourse to any other method of cure than merely reducing the Fire-place and the throat of the Chimney, or that part of it which lies immediately above the Fire-place, to a proper form, and just dimensions.

That my principles for constructing Fire-places are equally applicable to those which are designed for burning coal, as to those in which wood is burnt, has lately been abundantly proved by experiments made here in London; for of above an hundred and fifty Fire-places which have been altered in this city, under my direction, within these last two months, there is not one which has not answered perfectly well[1].—And by several experiments which have been made with great care, and with the assistance of thermometers, it has been demonstrated, that the saving of fuel, arising from these improvements of Fire-places, amounts in all cases to more than HALF, and in many cases to more than TWO THIRDS of the quantity formerly consumed.—Now as the alterations in Fire-places which are necessary may be made at a very trifling expence, as any kind of grate or stove may be made use of, and as no iron work, but merely a few bricks and some mortar, or a few small pieces of fire-stone, are required; the improvement in question is very important, when considered merely with a view to economy; but it should be remembered, that not only a great saving is made of fuel by the alterations proposed, but that rooms are made much more comfortable, and more salubrious;— that they may be more equally warmed, and more easily kept at any required temperature;—that all draughts of cold air from the doors and windows towards the Fire-place, which are so fatal to delicate constitutions, will be completely prevented;—that in consequence of the air being equally warm all over the room, or in all parts of it, it may be entirely changed with the greatest facility, and the room completely ventilated, when this air is become unfit for respiration, and this merely by throwing open for a moment a door opening into some passage from whence fresh air may be had, and the upper part of a window; or by opening the upper part of on window and the lower part of another, and as the operation of ventilating the room, even when it is done in the most complete manner, will never require the door and window to be open more than one minute; in this short time the walls of the room will not be sensibly cooled, and the fresh air which comes into the room will, in a very few minutes, be so completely warmed by these walls that the temperature of the room, though the air in it be perfectly changed, will be brought to be very nearly the same as it was before the ventilation.

Those who are acquainted with the principles of pneumatics, and know why the warm air in a room rushes out at an opening made for it at the top of a window when colder air from without is permitted to enter by the door, or by any other opening situated lower than the first, will see, that it would be quite impossible to ventilate a room in the complete and expeditious manner here described, where the air in a room is partially warmed, or hardly warmed at all, and where the walls of the room, remote from the fire, are constantly cold; which must always be the case where, in consequence of a strong current up the Chimney, streams of cold air are continually coming in through all the crevices of the doors and windows, and flowing into the Fire-place.

But although rooms, furnished with Fire-places constructed upon the principles here recommended, may be easily and most effectually ventilated, (and this is certainly a circumstance in favour of the proposed improvements,) yet such total ventilations will very seldom, if ever, be necessary. — As long as ANY FIRE is kept up in the room, there is so considerable a current of air up the Chimney, notwithstanding all the reduction that can be made in the size of its throat, that the continual change of air in the room which this current occasions will, generally, be found to be quite sufficient for keeping the air in the room sweet and wholesome; and indeed in rooms in which there is no open Fire-place, and consequently no current of air from the room setting up the Chimney, which is the case in Germany, and all the northern parts of Europe, where rooms are heated by stoves, whose Fire-places opening without are not supplied with the air necessary for the combustion of the fuel from the room; — and although in most of the rooms abroad, which are so heated, the windows and doors are double, and both are closed in the most exact manner possible, by slips of paper pasted over the crevices, or by slips of list or furr; yet when these rooms are tolerably large, and when they are not very much crowded by company, nor filled with a great many burning lamps or candles, the air in them is seldom so much injured as to become oppressive or unwholesome; and those who inhabit them show by their ruddy countenances, as well as by every other sign of perfect health, that they suffer no inconvenience whatever from their closeness. — There is frequently, it is true, an oppressiveness in the air of a room heated by a German stove, of which those who are not much accustomed to living in those rooms seldom fail to complain, and indeed with much reason; but this oppressiveness does not arise from the air of the room being injured by the respiration and perspiration of those who inhabit it; — it arises from a very different cause; — from a fault in the construction of German stoves in general, but which may be easily and most completely remedied, as I shall show more fully in another place. In the mean time, I would just observe here with regard to these stoves, that as they are often made of iron, and as this metal is a very good conductor of heat, some part of the stove in contact with the air of the room becomes so hot as to calcine or rather to ROAST the dust which lights upon it; which never can fail to produce a very disagreeable effect on the air of the room. And even when the stove is constructed of pantiles or pottery-ware, if any part of it in contact with the air of the room is suffered to become very hot, which seldom fails to be the case in German stoves constructed on the common principles, nearly the same effects will be found to be produced on the air as when the stove is made of iron, as I have very frequently had occasion to observe.

Though a room be closed in the most perfect manner possible, yet, as the quantity of air injured and rendered unfit for further use by the respiration of two or three persons in a few hours is very small, compared to the immense volume of air which a room of a moderate size contains; and as a large quantity of fresh air always enters the room, and an equal quantity of the warm air of the room is driven out of it every time the door is opened, there is much less danger of the air of a room becoming unwholesome for the want of ventilation than has been generally imagined; particularly in cold weather, when all the different causes which conspire to change the air of warmed rooms act with increased power and effect.

Those who have any doubts respecting the very great change of air or ventilation which takes place each time the door of a warm room is opened in cold weather, need only set the door of such a room wide open for a moment, and hold two lighted candles in the door-way, one near the top of the door, and the other near the bottom of it; the violence with which the flame of that above will be driven outwards, and that below inwards, by the two strong currents of air which, passing in opposite directions, rush in and out of the room at the same time, will be convinced that the change of air which actually takes place must be very considerable indeed; and these currents will be stronger, and consequently the change of air greater, in proportion as the difference is greater between the temperature of the air within the room and of that without. I have been more particular upon this subject,—the ventilation of warmed rooms which are constantly inhabited,—as I know that people in general in this country have great apprehensions of the bad consequences to health of living rooms in which there is not a continual influx of cold air from without. I am as much an advocate for a FREE CIRCULATION of air as any body, and always sleep in a bed without curtains on that account; but I am much inclined to think, that the currents of cold air which never fail to be produced in rooms heated by Fire-places constructed upon the common principle,— those partial heats on one side of the body, and the cold blasts on the other, so often felt in houses in this country, are infinitely more detrimental to health than the supposed closeness of the air in a room warmed more equally, and by a smaller fire.

All these advantages, attending the introduction of the improvements in Fire-places here recommended, are certainly important, and I do not know that they are counterbalanced by any one disadvantage whatsoever. The only complaints that I had ever heard made against them was, that they made the rooms TOO warm; but the remedy to this evil is so perfectly simple and obvious, that I should be almost afraid to mention it, less it might be considered as an insult to the understanding of the persons to whom such information should be given; for nothing surely can be conceived more perfectly ridiculous than the embarrassment of a person on account of the too great heat of his room, when it is in his power to diminish AT PLEASURE the fire by which it is warmed; and yet, strange as it may appear, this has sometimes happened!

Before I proceed to give directions for the construction of Fire-places, it will be proper to examine more carefully the Fire-places now in common use;—to point out their faults;— and to establish the principles upon which Fire-places ought to be constructed.

The great fault of all the open Fire-places, or Chimnies, for burning wood or coals in an open fire, now in common use, is, that they are much too large; or rather it is THE THROAT OF THE CHIM-NEY or the lower part of its open canal, in the neighbourhood of the mantle, and immediately over the fire, which is too large. This opening has hitherto been left larger than otherwise it probably would have been made, in order to give a passage to the Chimney-sweeper; but I shall show hereafter how a passage for the Chimney-sweeper may be contrived without leaving the throat of the Chimney of such enormous dimensions as to swallow up and devour all the warm air of the room, instead of merely giving a passage to the smoke and heated vapour which rise from the fire, for which last purpose alone it ought to be destined.

Were it my intention to treat my subject in a formal scientific manner, it would be doubtless be proper, and even necessary, to begin by explaining in the fullest manner, and upon the principles founded on the laws of nature, relative to the motions of elastic fluids, as far as they have been discovered and demonstrated, the causes of the ascent of smoke, and also to explain and illustrate upon the same principles, and even to measure, or estimate by cal-culations, the precise effects of all those mechanical aids which may be proposed for assisting it in its ascent, or rather for removing those obstacles which hinder its motion upwards;—but as it is my wish rather to write an useful practical treatise, than a learned dis-sertation, being more desirous to contribute in diffusing useful knowledge, by which the comforts and enjoyments of mankind may be increased, than to acquire the reputation of a philosopher among learned men, I shall endeavour to write in such a manner as to be easily understood BY THOSE WHO ARE MOST LIKELY TO PROF-IT BY THE INFORMATION I HAVE TO COMMUNICATE, and consequently most likely to assist in bringing into general use the improvements I recommend. This being premised, I shall proceed, without any further preface or introduction, to the investigation of the subject I have undertaken to treat.

As the immoderate size of the throats of Chimnies is the great fault of their construction, it is this fault which ought always to be first attended to in every attempt which is made to improve them; for however perfect the construction of a Fire-place may be in other respects, if the opening left for the passage of the smoke is larger than is necessary for that purpose, nothing can prevent the warm air of the room from escaping through it; and whenever this happens, there is not only an unnecessary loss of heat, but the warm air which leaves the room to go up the Chimney being replaced by cold air from without, the draughts of cold air, so often mentioned, cannot fail to be produced in the room, to the great annoyance of those who inhabit it. But although both these evils may be effectually remedied by reducing the throat of the Chimney to a proper size, yet in doing this several precautions will be necessary. And first of all, the throat of the Chimney should be in its proper place; that is to say, in that place in which it ought to be, in order that the ascent of the smoke may be most facilitated; for every means which can be employed for facilitating the ascent of the smoke in the Chimney must naturally tend to prevent the Chimney from smoking: now as the smoke and hot vapour which rise from a fire naturally tend UPWARDS, the proper place for the throat of the Chimney is evidently perpendicularly OVER THE FIRE.

But there is another circumstance to be attended to in determining the proper place for the throat of a Chimney, and that is, to ascertain its distance from the fire, or HOW FAR above the burning fuel it ought to be placed. In determining this point, there are many things to be considered, and several advantages and disadvantages to be weighed and balanced.

As the smoke and vapour which ascend from burning fuel rise in consequence of their being rarefied by heat, and made lighter than the air of the surrounding atmosphere; and as the degree of their rarefaction, and consequently their tendency to rise, is in proportion to the intensity of their heat; and further, as they are hotter near the fire than at a greater distance from it, it is clear that the nearer the throat of a Chimney is to the fire, the stronger will be, what is commonly called, its DRAUGHT, and the less danger there will be of its smoking. But on the other hand, when the draught of a Chimney is very strong, and particularly when this strong draught is occasioned by the throat of the Chimney being very near the fire, it may so happen that the draught of air into the fire may become so strong, as to cause the fuel to be consumed too rapidly. There are likewise several other inconveniences which would attend the placing of the throat of a Chimney VERY NEAR the burning fuel. In introducing the improvements proposed, in Chimnies already built, there can be no question in regard to the height of the throat of the Chimney, for its place will be determined by the height of the mantle. It can hardly be made lower than the mantle; and it ought always to be brought down as nearly upon the level with the bottom of it as possible. If the Chimney is apt to smoke, it will sometimes be necessary either to lower the mantle or to diminish the height of the opening of the Fire-place, by throwing over a flat arch, or putting in a straight piece of stone from one side of it to the other, or, which will be still more simple and easy in practice, building a wall of bricks, supported by a flat bar of iron, immediately under the mantle.

Nothing is so effectual to prevent Chimnies from smoking as diminishing the opening of the Fire-place in the manner here described, and lowering and diminishing the throat of the Chimney; and I have always found, except in the single instance already mentioned, that a perfect cure may be effected by THESE MEANS ALONE, even in the most desperate cases. It is true, that when the construction of the Chimney is very bad indeed, or its situation very unfavourable to the ascent of the smoke, and especially when both these disadvantages exist at the same time, it may sometimes be necessary to diminish the opening of the Fire-place, and particularly to lower it, and also to lower the throat of the Chimney, more than might be wished: but still I think this can produce no inconveniences to be compared with that greatest of all plagues, a smoking Chimney.

The position of the throat of a Chimney being determined, the next points to be ascertained are its size and form, and the manner in which it ought to be connected with the Fire-place below, and with the open canal of the Chimney above.

But as these investigations are intimately connected with those which relate to the form proper to be given to the Fire-place itself, we must consider them all together.

That these inquiries may be pursued with due method, and that the conclusions drawn from them may be clear and satisfactory, it will be necessary to consider, first, what the objects are which ought principally to be had in view in the construction of a Fire-place; and secondly, to see how these objects can best be attained.

Now the design of a Chimney Fire being simply to warm a room, it is necessary, first of all, to contrive matters so that the room shall be actually warmed; secondly, that it be warmed with the smallest expence of fuel possible; and, thirdly, that in warming it, the air of the room be preserved perfectly pure, and fit for respiration, and free from smoke and all disagreeable smells.

In order to take measures with certainty for warming a room by means of an open Chimney Fire, it will be necessary to consider HOW, or in WHAT MANNER, such a Fire communicates heat to a room. This question may perhaps, at the first view of it, appear to be superfluous and trifling, but a more careful examination of the matter will show it to be highly deserving of the most attentive investigation.

To determine in what manner a room is heated by an open Chimney Fire, it will be necessary first of all to find out, UNDER WHAT FORM the heat generated in the combustion of the fuel exists, and then to see how it is communicated to those bodies which are heated by it.

In regard to the first of these subjects of inquiry, it is quite certain that the heat which is generated in the combustion of the fuel exists under TWO perfectly distinct and very different forms. One part of it is COMBINED with the smoke, vapour, and heated air which rise from the burning fuel, and goes off with them into the upper regions of the atmosphere; while the other part, which appears to be UNCOMBINED, or, as some ingenious philosophers have supposed, combined only with light, is sent off from the fire in rays in all possible directions.

With respect to the second subject of inquiry; namely, how this heat, existing under these two different forms, is communicated to other bodies; it is highly probable that the combined heat can only be communicated to other bodies by ACTUAL CONTACT with the body with which it is combined; and with regard to the rays which are sent off by burning fuel, it is certain that THEY communicate or generate heat only WHEN and WHERE they are stopped or absorbed. In passing through air, which is transparent, they certainly do not communicate any heat to it; and it seems highly probable that they do not communicate heat to solid bodies by which they are reflected.

In these respects they seem to bear a great resemblance to the solar rays. But in order not to distract the attention of my reader, or carry him too far away from the subject more immediately under consideration, I must not enter too deeply into these inquiries respecting the nature and properties of what has been called RADIANT HEAT. It is certainly a most curious subject of philosophical investigation, but more time would be required to do it justice than we now have to spare. We must therefore content ourselves with such a partial examination of it as will be sufficient for our present purpose.

A question which naturally presents itself here is. What proportion does the radiant heat bear to the combined heat? — Though that point has not yet been determined with any considerable degree of precision, it is, however, quite certain, that the quantity of heat which goes off combined with the smoke, vapour, and heated air is much more considerable, perhaps three of four times greater at least, than that which is sent off from the fire in rays. — And yet, small as the quantity is of this radiant heat, it is the only part of the heat generated in the combustion of fuel burnt in an open Fire-place which is ever employed, or which can ever be employed, in heating a room.

The whole of the combined heat escapes by the Chimney, and is totally lost; and, indeed, no part of it could ever be brought into a room from an open Fire-place, without bringing along with it the smoke with which it is combined; which, of course, would render it impossible for the room to be inhabited. There is, however, one method by which combining heat, and even that which arises from an open Fire-place, may be made to assist in warming a room; and that is by making it pass through something analogous to a German stove, placed in the Chimney above the fire. — But of this contrivance I shall take occasion to treat more fully hereafter; in the mean time I shall continue to investigate the properties of open Chimney Fire-places, constructed upon the most simple principles, such as are now in common use; and shall endeavour to point out and explain all those improvements of which THEY appear to me to be capable. When fuel is burnt in Fire-places upon this simple construction, where the smoke escapes immediately by the open canal of the Chimney, it is quite evident that all the combined heat must of necessity be lost; and as it is the radiant heat alone which can be employed in heating a room, it becomes an object of much importance to determine how the greatest quantity of it may be generated in the combustion of the fuel, and how the greatest proportion possible of that generated may be brought into the room.

Now the quantity of radiant heat generated in the combustion of a given quantity of any kind of fuel depends very much upon the management of the fire, or upon the manner in which the fuel is consumed. When the fire burns bright, much radiant heat will be sent off from it; but when it is SMOTHERED UP, very little will be generated; and indeed very little combined heat, that can be employed to any useful purpose: most of the heat produced will be immediately EXPENDED in giving elasticity to a thick dense vapour or smoke which will be seen rising from the fire; — and the combustion being very incomplete, a great part of the inflammable matter of the fuel being merely rarefied and driven up the Chimney without being inflamed, the fuel will be wasted to little purpose. And hence it appears of how much importance it is, whether it be considered with a view to economy, or to cleanliness, comfort, and elegance, to pay due attention to the management of a Chimney Fire.

Nothing can be more perfectly void of common sense, and wasteful and slovenly at the same time, than the manner in which Chimney Fires, and particularly where coals are burned, are commonly managed by servants. They throw on a load of coals at once, through which the flame is hours in making its way; and frequently it is not without much trouble that the fire is prevented from going quite out. During this time no heat is communicated to the room; and what is still worse, the throat of the Chimney being occupied merely by a heavy dense vapour, not possessed of any considerable degree of heat, and consequently not having much elasticity, the warm air of the room finds less difficulty in forcing its way up the Chimney and escaping, than when the fire burns bright;—and it happens not unfrequently, especially in Chimneys and Fire-places ill constructed, that this current of warm air from the room which presses into the Chimney, crossing upon the current of heavy smoke which rises slowly from the fire, obstructs it in its ascent, and beats it back into the room; hence it is that Chimnies so often smoke when too large a quantity of fresh coals is put upon the fire. So many coals should never be put on the fire at once as to prevent the free passage of the flame between them. In short, a fire should never be smothered; and when proper attention is paid to the quantity of coals put on, there will be very little use for the poker; and this circumstance will contribute very much to cleanliness, and to the preservation of furniture.

Those who have feeling enough to be made miserable by any thing careless, slovenly, and wasteful which happens under their eyes,—who know what comfort is, and consequence are worthy of the enjoyments of a CLEAN HEARTH and a CHEERFUL FIRE, should really either take the trouble themselves to manage their fires, (which, indeed, would rather be an amusement to them than a trouble,) or they should instruct their servants to manage them better.

But to return to the subject more immediately under consideration. As we have seen what is necessary to the production or generation of radiant heat, it remains to determine how the greatest proportion of that generated and sent off from the fire in all directions may be made to enter the room, and assist in warming it. How as the rays which are thrown off from burning fuel have this property in common with light, that they generate heat only WHEN and WHERE they are stopped or absorbed, and also in being capable of being reflected WITHOUT GENERATING at the surfaces of various bodies, the knowledge of these properties will enable us to take measures, with the utmost certainty, for producing the effect required,—that is to say, for bringing as much radiant heat as possible into the room.

This must be done, first, by causing as many as possible of the rays, as they are sent off from the fire in straight lines, to come DIRECTLY into the room; which can only be effected by bringing the fire as far forward as possible, and leaving the opening of the Fireplace as wide and as high as can be done without inconveniences; and secondly, by making the sides and back of the Fire-place of such form, and constructing them of such materials, as to cause the direct rays from the fire, which strike against them, to be sent into the room BY REFLECTION in the greatest abundance.

Now it will be found, upon examination, that the best form for the vertical sides of a Fire-place, or the COVINGS, (as they are called,) is that of an upright plane, making an angle with the plane of the back of the Fire-place, of about 135 degrees. — According to the present construction of Chimnies this angle is 90 degrees, or forms a right angle; but as in this case the two sides or covings of the Fire-place (AC, BD, Fig. 1.) are parallel to each other, it is evident that they are very ill contrived for throwing into the room by reflection the rays from the fire which fall on them.

To have a clear and perfect idea of the alterations I propose in the forms of Fire-places, the reader need only observe, that, whereas the backs of Fire-places, as they are now commonly constructed, are as wide as the opening of the Fire-place in front, and the sides of it are of course perpendicular to it, and parallel to each other, — in the Fire-places I recommend, the back (i k, Fig. 3) is only about one-third of the width of the opening of the Fire-place in front (a,b), and consequently that the two sides of covings of the Fire-place (a i and b k), instead of being perpendicular to the back, are inclined to it at an angle of about 135 degrees; and in consequence of this position, instead of being parallel to each other, each of them presents an oblique front towards the opening of the Chimney, by means of which the rays which they reflect are thrown into the room. A bare inspection of the annexed drawings (Fig. 1. and Fig. 3.) will render this matter perfectly clear and intelligible.

In regard to the materials which it will be most advantageous to employ in the construction of Fire-places, so much light has, I flatter myself, already been thrown on the subject we are investigating, and the principles adopted have been established on such clear and obvious facts, that no great difficulty will attend the determination of that point. — As the object in view is to bring radiant heat into the room, it is clear that that material is best for the construction of a Fire-place which reflects the most, or which ABSORBS THE LEAST of it; for that heat which is ABSORBED cannot be REFLECTED — Now as bodies which absorb radiant heat are necessarily heated in consequence of that absorption, to discover which of the various materials that can be employed for constructing Fire-places are best adapted for that purpose, we have only to find out by an experiment, very easy to be made, what bodies acquire LEAST HEAT when exposed to the direct rays of a clear fire;—for those which are least heated, evidently absorb the least, and consequently reflect the most radiant heat. And hence it appears that iron, and, in general, metals of all kinds, which are well known to GROW VERY HOT when exposed to the rays projected by burning fuel, are to be reckoned among the VERY WORST materials that it is possible to employ in the construction of Fire-places.

The best materials I have hitherto been able to discover are firestone, and common bricks and mortar. Both these materials are, fortunately, very cheap; and as to their comparative merits, I hardly know to which of them the preference ought to be given.

When bricks are used they should be covered with a thin coating of plaster, which, when it is become perfectly dry, should be whitewashed. The fire-stone should likewise be white washed, when that is used; and every part of the Fire-place, which is not exposed to being soiled and made black by the smoke, should be kept as white and clean as possible. As WHITE reflects more heat, as well as more light than any other colour, it ought always to be preferred for the inside of a Chimney Fire-place, and BLACK, which reflects neither light nor heat should be most avoided.

I am well aware how much the opinion I have have ventured to give, respecting the unfitness of iron and other metals to be employed in the construction of open Fire-places, differs from the opinion generally received upon that subject;—and I even know that the very reason which, according to my ideas of the matter, renders them totally unfit for the purpose, is commonly assigned for making use of them, namely, that they soon grow very hot. But I would beg leave to ask what advantage is derived from heating them?

I have shown the disadvantage of it, namely, that the quantity of radiant heat thrown into the room is diminished;—and it is easy to show that almost the whole of that absorbed by the metal is ultimately carried up the Chimney by the air, which, coming into contact with this hot metal, is heated and rarefied by it, and forcing its way upwards, goes off with the smoke; and as no current of air ever sets from any part of the opening of a Fire-place into the room, it is impossible to conceive how the heat existing in the metal composing any part of the apparatus of the Fire-place, and situated within its cavity, can come, or be brought into the room.

This difficulty may be in part removed, by supposing, what indeed seems to be true in a certain degree, that the heated metal sends off rays, the heat it acquires from the fire, even when it is not heated red hot; but still, as it never can be admitted that the heat, absorbed by the metal and afterwards thrown off by it in rays, is INCREASED by this operation, nothing can be gained by it; and as much must necessary be lost in consequence of the great quantity of heat communicated by the hot metal to the air in contact with it, which, as has already been shown, always makes its way up the Chimney, and flies off into the atmosphere, the loss of heat attending the use of it is too evident to require being farther insisted on.

There is, however, in Chimney Fire-places destined for burning coals, one essential part, the grate, which cannot well be made of any thing else but iron; but there is no necessity whatever for that immense quantity of iron which surrounds grates as they are now commonly constructed and fitted up, and which not only renders them very expensive, but injures very essentially the Fire-place. If it should be necessary to diminish the opening of a large Chimney in order to prevent its smoking, it is much more simple, economical, and better in all respects, to do this with marble, fire-stone, or even with bricks and mortar, than to make use of iron, which, as has already been shown, is the very worst material that can possibly be employed for that purpose; and as to registers, they not only are quite unnecessary, where the throat of a Chimney is properly constructed, and of proper dimensions, but in that case would do much harm. If they act at all, it must be by opposing their flat surfaces to the current of rising smoke in a manner which cannot fail to embarrass and impede its motion. But we have shown that the passage of the smoke through the throat of a Chimney ought to be facilitated as much as possible, in order that it may be enabled to pass by a small aperture.

Register-stoves have often been found to be of use, but it is because the great fault of all Fire-places constructed upon the common principles being the enormous dimensions of the throat of the Chimney, this fault has been in some measure corrected by them; but I will venture to affirm, that there never was a Fire-place so corrected that would not have been much more improved, and with infinitely less expence, by the alterations here recommended, and which will be more particularly explained in the next Chapter.

CHAPTER. II.

Practical directions designed for the use of workmen, showing how they are to proceed in making the alterations necessary to improve chimney fire-places, and effectually to cure smoking chimnies.

All Chimney Fire-places, without exception, whether they are designed for burning wood or coals, and even those which do not smoke, as well as those which do, may be greatly improved by making the alterations in them here recommended; for it is by no means MERELY to prevent Chimnies from smoking that these improvements are recommended, but it is also to make them better in all other respects as Fire-places; and when the alterations proposed are properly executed, which may be very easily be done with the assistance of the following plain and simple directions, the Chimnies will never fail to answer, I will venture to say, even beyond expectation. The room will be heated much more equally and more pleasantly with LESS THAN HALF THE FUEL used before, the fire will be more cheerful and more agreeable; and the general appearance of the Fire-place more neat and elegant, and the Chimney WILL NEVER SMOKE.

The advantages which are derived from mechanical inventions and contrivances are, I know, frequently accompanied by disadvantages which it is not always possible to avoid; but in the case in question, I can say with truth, that I know of no disadvantage whatever that attends the Fire-places constructed upon the principles here recommended. —But to proceed in giving directions for the construction of these Fire-places.

That what I have to offer on this subject may be the more easily understood, it will be proper to begin by explaining the precise meaning of all those technical words and expressions which I may find it necessary or convenient to use.

By the THROAT of a Chimney, I mean the lower extremity of its canal, where it unites with the upper part of its open Fire-place. — This throat is commonly found about a foot above the level of the lower part of the mantle, and it is sometimes contracted to a smaller size than the rest of the canal of the Chimney, and sometimes not.

Fig. 5. shows the section of a Chimney on the common construction, in which d e is the throat.

Fig. 6. shows the section of the same Chimney altered and improved, in which d i is the reduced throat.

The BREAST of a Chimney, is that part of it which is immediately behind the mantle.—It is the wall which forms the entrance from below into the throat of the Chimney in front, or towards the room.—It is opposite to the upper extremity of the back of the open Fire-place, and parallel to it; in short it may said to be the back part of the mantle itself.—In the figures 5 and 6, it is marked by the letter d. The WIDTH of the throat of Chimney (d e fig. 5, and d i fig. 6,) is taken from the breast of the Chimney to the back, and its LENGTH is taken at right angles to its width, or in a line parallel to the mantle (a fig. 5. and 6.).

Before I proceed to give particular directions respecting the exact forms and dimensions of the different parts of a Fire-place, it may be useful to make such general an practical observations upon the subject as can be clearly understood without the assistance of drawings; for the more complete the knowledge of any subject is which can be acquired without drawings, the more easy will it be to understand the drawings when it becomes necessary to have recourse to them.

The bringing forward of the Fire into the room, or rather bringing it nearer to the front of the opening of the Fire-place;—and the diminishing of the throat of the Chimney, being two objects principally had in view in the alterations in Fire-places here recommended, it is evident that both these may be attained merely by bringing forward the back of the Chimney. —The only question therefore is, how far it should be brought forward?—The answer is short, and easy to be understood;—bring it forward as far as possible, without diminishing too much the passage which must be left for the smoke. Now as this passage, which, in its narrowest part, I have called the THROAT OF THE CHIMNEY, ought, for reasons which are fully explained in the foregoing Chapter, to be immediately, or perpendicularly over the Fire, it is evident that the back of the Chimney must always be built perfectly upright.—To determine therefore the place for the new back, or how far precisely it ought to be brought forward, nothing more is necessary than to ascertain how wide the throat of the Chimney ought to be left, or what space must be left, between the top of the breast of the Chimney, where the upright canal of the Chimney begins, and the new back of the Fire-place carried up perpendicularly to that height.

In the course of my numerous experiments upon Chimnies, I have taken much pains to determine the width proper to be given to this passage, and I have found, that, when the back of the Fire-place is of a proper width, the best width for the throat of a Chimney, when the Chimney and the Fire-place are at the usual form and size, is FOUR INCHES.—Three inches might sometimes answer, especially where the Fire-place is very small, and the Chimney good, and well situated: but as it is always of much importance to prevent those accidental puffs of smoke which are sometimes thrown into rooms by the carelessness of servants in putting on suddenly too many coals at once upon the fire, and as I found these accidents sometimes happened when the throats of Chimneys were made very narrow, I found that, upon the whole, all circumstances being well considered, and advantages and disadvantages compared and balanced, FOUR INCHES is the best width that can be given to the throat of a chimney; and this, whether the Fire-place be destined to burn wood, coals, turf, or any other fuel commonly used for heating rooms by an open fire.

In Fire-places destined for heating very large halls, and where very great fires are kept up, the throat of the Chimney may, if it should be thought necessary, be made four inches and an half, or five inches wide;—but I have frequently made Fire-places for halls which have answered perfectly well where the throats of the Chimnies have not been wider than four inches.

It may perhaps appear extraordinary, upon the first view of the matter, that Fire-places of such different sizes should all require the throat of the Chimney to be of the same width; but when it is considered that the CAPACITY of the throat of a Chimney does not depend on its width alone, but on its width and LENGTH taken together; and that in large Fire-places, the width of the back, and consequently the length of the throat of the Chimney, is greater than in those which are smaller, this difficulty vanishes.

And this leads us to consider another important point respecting open Fire-places, and that is, the width which it will, in each case, be proper to give to the back. — In Fire-places as they are now commonly constructed, the back is of equal width with the opening of the Fire-place in front; — but this construction is faulty on two accounts. — First, in a Fire-place, so constructed, the sides of the Fire-place, or COVINGS, as they are called, are parallel to each other, and consequently ill-contrived to throw out into the room the heat they receive from the fire in the form of rays; — and secondly, the large open corners which are formed by making the back as wide as the opening of the Fire-place in front occasion eddies of wind, which frequently disturb the fire, and embarrass the smoke in its ascent in such a manner as often to bring it into the room. — Both these defects may be entirely remedied by diminishing the width of the back of the Fire-place. — The width which, in most cases, it will be best to give it, is ONE THIRD of the width of the opening of the Fire-place in front. — But it is not absolutely necessary to conform rigorously to this decision, nor will it always be possible. — It will frequently happen that the back of a Chimney must be made wider than, according to the rule here given, it ought to be. — This may be, either to accommodate the Fire-place to a stove, which being already on hand, must, to avoid the expense of purchasing a new one, be employed; or for other reasons; — and any small deviation from the general rule will be attended with no considerable inconvenience. — It will always be best, however, to conform to it as far as circumstances will allow.

Where a Chimney is designed for warming a room of a middling size, and where the thickness of the wall of the Chimney in front, measured from the front of the mantle to the breast of the Chimney, is nine inches, I should set off four inches more for the width of the throat of the Chimney, which, supposing the back of the Chimney to be built upright, as it always ought to be, will give thirteen inches for the depth of the Fire-place, measured upon the hearth, from the opening of the Fire-place in front, to the back. — In this case thirteen inches would be a good size for the width of the back; and three times thirteen inches, or thirty-nine inches, for the width of the opening of the Fire-place in front; and the angle made by the back of the Fire-place and the sides of it, or covings, would be just 135 degrees, which is the best position they can have for throwing heat into the room.

But I will suppose that in altering such a Chimney it is found necessary, in order to accommodate the Fire-place to a grate or stove already on hand, to make the Fire-place sixteen inches wide. — In that case, I should merely increase the width of the back, to the dimensions required, without altering the depth of the Chimney, or increasing the width of the opening of the Chimney in front. — The covings, it is true, would be somewhat reduced in their width, by this alteration; and their position with respect to the plane of the back of the Chimney would be a little changed; but these alterations would produce no bad effects of any considerable consequence, and would be much less likely to injure the Fire-place, than an attempt to bring the proportions of its parts nearer to the standard, by increasing the depth of the Chimney, and the width of its opening in front; — or than an attempt to preserve that particular obliquity of the covings which is recommended as the best, (135 degrees,) by increasing the width of the opening of the Fire-place, without increasing its depth.

In order to illustrate this subject more fully, we will suppose one case more. — We will suppose that in the Chimney which is to be altered, the width of the Fire-place in front is either wider or narrower than it ought to be, in order that the different parts of the Fire-place, after it is altered, may be of the proper dimensions. In this case, I should determine the depth of the Fire-place, and the width of the back of it, without any regard to the width of the opening of the Fire-place in front; and when this is done, if the opening of Fire-place should be only two or three inches too wide, that is to say, only two or three inches wider than is necessary in order that the covings may be brought into their proper position with respect to the back, I should not alter the width of this opening, but should accommodate the covings to this width, by increasing their breadth, and increasing the angle they make with the back of the Fire-place; but if the opening of the Fire-place should be more than three inches too wide; — I should reduce it to the proper width by slips of stone, or by bricks and mortar.

When the width of the opening of the Fire-place, in front, is very
great, compared with the depth of the Fire-place, and with the
width of the back, the covings in that case being very wide, and
consequently very oblique, and the Fire-place very shallow, any
sudden motion of the air in front of the Fire-place, (that motion, for
instance, which would be occasioned by the clothes of a woman
passing hastily before the fire, and very near it,) would be apt to
cause eddies in the air, WITHIN THE OPENING OF THE FIRE-
PLACE, by which puffs of smoke might easily be brought into the
room. Should the opening of the Chimney be too narrow, which
however will very seldom be found to be the case, it will, in general,
be advisable to let it remain as it is, and to accommodate the covings
to it, rather to attempt to increase its width, which would be attended with a good deal of trouble, and probably a considerable expence.

From all that has been said it is evident, that the points of the
greatest importance, and which ought most particularly to be attended to, in altering Fire-places upon the principles here recommended, are, the bringing forward the back to its proper place, and
making it of a proper width. — But it is time that I should mention
another matter upon which it is probable that my reader is already
impatient to receive information. — Provision must be made for the
passage of the Chimney-sweeper up the Chimney. — This may easily
be done in the following manner: — In building up the new back of
the Fire-place; when this wall, (which need never be more than the
width of a single brick in thickness,) is brought up so high that there
remains no more than about ten or eleven inches between what is
then the top of it, and the inside of the mantle, or lower extremity of
the breast of the Chimney, an opening, or door-way, eleven or
twelve inches wide, must be begun in the middle of the back, and
continued quite to the top of it, which, according to the height to
which it will commonly be necessary to carry up the back, will
make the opening about twelve or fourteen inches high; which will
be quite sufficient to allow the Chimney-sweeper to pass. When the
Fire-place is finished, this door-way is to be closed by a few bricks,
by a tile, or a fit piece of stone, placed in it, dry, or without mortar,
and confined in its place by means of a rabbet made for that purpose in the brick-work. — As often as the Chimney is swept, the

Chimney-sweeper takes down this temporary wall, which is very easily done, and when he has finished his work, he puts it again into its place. — The annexed drawing (No. 6.) will give a clear idea of this contrivance; and the experience I have had of it has proved that it answers perfectly well the purpose for which it is designed.

I observed above, that the new back, which it will always be found necessary to build in order to bring the fire sufficiently forward, in altering a Chimney constructed on the common principles, need never be thicker than the width of a common brick. — I may say the same of the thickness necessary to be given to the new sides, or covings, of the Chimney; or if the new back and covings are constructed of stone, one inch and three quarters, or two inches in thickness will be sufficient. — Care should be taken in building up these new walls to unite the back to the covings in a solid manner.

Whether the new back and covings are constructed of stone, or built of bricks, the space between them, and the old back and covings of the Chimney ought to be filled up, to give greater solidity to the structure. — This may be done with loose rubbish, or pieces of broken bricks, or stones provided the work be strengthened by a few layers or courses of bricks laid in mortar; but it will be indispensably necessary to finish the work, where these new walls end, that is to say, at the top of the throat of the Chimney, where it ends abruptly in the open canal of the Chimney by a horizontal course of bricks well secured with mortar. — This course of bricks will be upon a level with the top of the door-way left for the Chimney-sweeper.

From these descriptions it is clear that where the throat of the Chimney has an end, that is to say, where it enters into the lower part of the open canal of the Chimney, THERE the three walls which form the two covings and the back of the Fire-place all end abruptly. — It is of much importance that they should end in this manner; for were they to be sloped outward and raised in such a manner as to swell out the upper extremity of the throat of the Chimney in the form of a trumpet, and increase it by degrees to the size of the canal of the Chimney, this manner of uniting the lower extremity of the canal of the Chimney with the throat would tend to assist the winds which may attempt to blow down the Chimney, in forcing their way through the throat, and throwing the smoke backward into the room; but when the throat of the Chimney ends abruptly, and the ends of the new walls form a flat horizontal surface, it will be much more difficult for any wind from above, to find, and force its way through the narrow passage of the throat of the Chimney.

As the two walls which form the new covings of the Chimney are not parallel to each other; but inclined, presenting an oblique surface towards the front of the Chimney, and as they are built perfectly upright and quite flat, from the hearth to the top of the throat, where they end, it is evident that an horizontal section of the throat will not be an oblong square; but its deviation from that form is a matter of no consequence; and no attempts should ever be made, by twisting the covings above, where they approach the breast of the Chimney, to bring it to that form. — All twists, bends, prominences, excavations, and other irregularities of form, in the covings of a Chimney, never fail to produce eddies in the current of air which is continually passing into, and through an open Fire-place in which a fire is burning; — and all such eddies disturb, either the fire, or the ascending currents of smoke, or both; and not unfrequently cause the smoke to be thrown back into the room. — Hence it appears, that the covings of Chimneys should never be made circular, or in the form of any other curve; but always quite flat.

For the same reason, that is to say, to prevent eddies, the breast of the Chimney, which forms that side of the throat that is in front, or nearest to the room, should be nearly cleaned off, and its surface made quite regular and smooth.

This may easily be done by covering it with a coat of plaster, which may be made thicker or thinner in different parts as may be necessary in order to bring the breast of the Chimney to be of the proper form.

With regard to the form of the breast of a Chimney, this is a matter of very great importance, and which ought always to be particularly attended to. — The worst form it can have is that of a vertical plane, or upright flat; — and next to this the worst form is an inclined plane. — Both these forms cause the current of warm air from the room, which will, in spite of every precaution, sometimes find its way into the Chimney, to cross upon the current of smoke, which rises from the fire, in a manner most likely to embarrass it in its ascent, and drive it back. —The inclined plane which is formed by a flat register placed in the throat of a Chimney produces the same effects; and this is one reason, among many others, which have induced me to disapprove of register stoves.

The current of air, which, passing under the mantle, gets into the Chimney, should be made GRADUALLY TO BEND ITS COURSE UPWARDS, by which means it will be QUIETLY with the ascending current of smoke, and will be less likely to check it, or force it back into the room.—Now this may be effected with the greatest ease and certainty, merely by ROUNDING OFF the breast of the Chimney or back part of the mantle, instead of leaving it flat, or full of holes and corners; and this of course ought always to be done.

I have hitherto given no precise directions in regard to the height to which the new back and covings ought to be carried:— This will depend not only on the height of the mantle, but also, and more especially, on the height of the breast of the Chimney, or of that part of the Chimney where the breast ends and the upright canal begins.—The back and covings must rise a few inches, five or six for instance, higher than this part, otherwise the throat of the Chimney will not be properly formed:—but I know of no advantages that would be gained by carrying them up still higher.

I mentioned above, that the space between the walls which form the new back and covings, and the old back and sides of the Fireplace, should be filled up:—but this must not be understood to apply to the space between the wall of dry bricks, or the tile which closes the passage for the Chimney-sweeper, and the old back of the Chimney; for that space must be left void, otherwise, though this tile (which at most will not be more than two inches in thickness,) were taken away, there would not be any room sufficient for him to pass.

In forming this door-way, the best method of proceeding is to place the tile or flat piece of stone destined for closing it, in its proper place; and to build round it, or rather by the sides of it; taking care not to bring any mortar near it, in order that it may be easily removed when the door-way is finished. — With regard to the rabbet which should be made in the door-way to receive it and fix it more firmly in its place, this may either be formed at the same time when the door-way is built, or it may be made after it is finished, by attaching to its bottom and sides, with strong mortar, pieces of thin roof tiles. Such as are about half an inch in thickness will be best for this use; if they are thicker, they will diminish too much the opening of the door-way, and will likewise be more liable to be torn away by the Chimney-sweeper in passing up and down the Chimney.

It will hardly be necessary for me to add, that the tile, or flat stone, or wall of dry bricks, which is used for closing up the door-way, must be of sufficient height to reach quite up to a level with the top of the walls which form the new back and covings of the Chimnies.

I ought, perhaps, to apologize for having been so very particular in these description and explanations, but it must be remembered that this chapter is written principally for the information of those who, having had few opportunities of employing their attention in abstruse philosophical researches, are not sufficiently practised in these intricate investigations, to seize, with facility, new ideas; — and consequently, that I have frequently been obliged TO LABOUR to make myself understood.

I have only to express my wishes that my reader may not be more FATIGUED with this labour than I have been; — for we shall them most certainly be satisfied with each other. — But to return once more to the charge.

There is one important circumstance respecting Chimney Fire-places, destined for burning coals, which still remains to be farther examined; — and that is the Grate.

Although there are few grates that may not be used in Chimneys constructed or altered upon the principles here recommended, yet they are not, by any means, all equally well adapted for that purpose. — Those whose construction is the most simple, and which of course are the cheapest, are beyond comparison the best, ON ALL ACCOUNTS. — Nothing being wanted in these Chimnies but merely a grate for containing the coals, and in which they will burn with a clear fire;—and all additional apparatus being, not only useless, but very pernicious, all complicated and expensive grates should be laid aside, and such as more simple substituted in the room of them.—And in the choice of a grate, as in every thing else, BEAUTY and ELEGANCE may easily be united with the MOST PERFECT SIMPLICITY.—Indeed they are incompatible with every thing else.

In placing the grate, the thing principally to be attended to is, to make the back of it coincide with the back of the Fire-place;— but as many of the grates now in common use will be found to be too large, when the Fire-places are altered and improved, it will be necessary to diminish their capacities by filling them up at the back and the sides with pieces of fire-stone. When this is done, it is the front of the flat piece of fire-stone which is made to form a new back to the grate, which must be made to coincide with, and make part of the back, of the Fire-place.— But in diminishing the capacities of grates with pieces of fire-stone, care must be taken not to make them TOO NARROW.

The proper width for grates destined for rooms of a middling size will be from six to eight inches, and their length may be diminished more or less, according as the room is heated with more or less difficulty, or as the weather is more or less severe. —But where the width of a grate is not more than five inches, it will be very difficult to prevent the fire from going out.

It goes out for the same reason that a live coal from the grate that falls upon the hearth soon ceases to be red hot;—it is cooled by the surrounding cold air of the atmosphere.— The knowledge of the cause which produces this effect is important, as it indicates the means which may be used for preventing it. —But of this subject I shall treat more fully hereafter.

It frequently happens that the iron backs of grates are not vertical, or upright, but inclined backwards.—When these grates are so much too wide as to render it necessary to fill them up behind with fire-stone, the inclination of the back will be of little consequence; for by making the piece of stone with which the width of the grate is to be diminished in the form of a wedge, or thicker above than below, the front of this stone, which in effect will become the back of the grate, may be made perfectly vertical; and the iron back of the grate being hid in the solid work of the back of the Fire-place, will produce no effect whatever; but if the grate be already so narrow as not to admit of any diminution of its width, in that case it will be best to take away the iron back of the grate entirely, and fixing the grate firmly in the brick-work, cause the back of the Fire-place to serve as a back to the grate. — This I have very frequently done, and have always found it to answer perfectly well.

Where it is necessary that the fire in a grate should be very small, it will be best, in reducing the grate with fire-stone, to bring its cavity, destined for containing the fuel, to the form of one half of a hollow hemisphere; the two semicircular openings being one above, to receive the coals, and the other in front, or towards the bars of the grate; for when the coals are burnt in such a confined space, and surrounded on all sides, except in the front and above, by fire-stone, (a substance peculiarly well adapted for confining heat,) the heat of the fire will be concentrated, and the cold air of the atmosphere being kept at a distance, a much smaller quantity of coals will burn, than could possibly be made to burn in a grate where they would be more exposed to be cooled by the surrounding air, or to have their heat carried off by being in contact with iron, or with any other substance through which heat passes with greater facility than through fire-stone.

Being persuaded that if the improvements in Chimney Fire-places here recommended should be generally adopted, (which I cannot help flattering myself will be the case,) that it will become necessary to reduce, very considerably, the sizes of grates, I was desirous of showing how this may, with the greatest safety and facility, be done.

Where grates, which are designed for rooms of a middling size, are longer than 14 or 15 inches, it will always be best, not merely to diminish their lengths, by filling them up at their two ends with fire-stone, but, forming the back of the Chimney of a proper width, without paying any regard to the length of the grate, to carry the covings through the two ends of the grate in such a manner as to conceal them, or at least to conceal the back corners of them in the walls of the covings.

I cannot help flattering myself that the directions here given in regard to the alterations which it may be necessary to make in Fire-places, in order to introduce the improvements proposed, will be found to be so perfectly plain and intelligible that no one who reads them will be at any loss respecting the manner in which the work is to be performed; — but as order and arrangement tend much to facilitate all mechanical operations, I shall here give a few short directions respecting the manner of LAYING OUT THE WORK, which may be found useful, and particularly to gentlemen who may undertake to be their own architects, in ordering and directing the alterations to be made for the improvement of their Fire-places.

Directions for laying out the Work.

If there be a grate in the Chimney which is to be altered, it will always be best to take it away; and when this is done, the rubbish must be removed, and the hearth swept perfectly clean.

Suppose the annexed figure No. 1. to represent the ground plan of such a Fire-place; A B being the opening of it in front, A C and B D the two sides or covings, and C D the back.

Figure 2. shows the elevation of this Fire-place.

First draw a strait line, with chalk, or with a lead pencil, upon the hearth, from one jamb to the other,—even with the front of the jambs. The dotted line A B, figure 3, may represent this line.

From the middle C of this line, (A B) another line c d, is to be drawn perpendicular to it, across the hearth, to the middle d, of the back of the Chimney.

A person must now stand upright in the Chimney, with his back to the back of the Chimney, and hold a plumb-line to the middle of the upper part of the breast of the Chimney (d, fig. 5,) or where the canal of the Chimney begins to rise perpendicularly;— taking care to place the line above in such a manner that the plumb may fall on the line c d, draw on the hearth from the middle of the opening of the Chimney in front to the middle of the back, and an assistant must mark the precise place e, on that line where the plumb falls.

This being done, and the person in the Chimney having quitted his station, four inches are to be set off the line c d, from e, towards d; and the point f, where these four inches end, (which must be marked with chalk, or with a pencil,) will show how far the new back is to be brought forward.

Through f, draw the line g h, parallel to the line A B, and this line g h will show the direction of the new back, or the ground line upon which it is to be built.

The line c f will show the depth of the new Fire-place; and if it should happen that c f is equal to about ONE-THIRD of the line A B; and if the grate can be accommodated to the Fire-place instead of its being necessary to accommodate the Fire-place to the grate, in that case, half the length of the line c f, is to be set off from f on the line g f h, on one side to k, and on the other to i, and the line i k will show the ground line of the fore part of the back of the Chimney.

In all cases where the width of the opening of the Fire-place in front (A B) happens to be not greater, or not more than two or three inches greater than THREE TIMES the width of the new back of the Chimney (i k), this opening may be left, and lines drawn from i to A, and from k to B, will show the width and position of the front of the new covings;—but when the opening of the Fire-place in front is still wider, it must be reduced; which is to be done in the following manner:

From c, the middle of the line A B, c a and c b, must be set off equal to the width of the back (i k), added to half its width (f i), and lines drawn from i to a, and from k to b, will show the ground plan of the fronts of the new covings.

When this is done, nothing more will be necessary than to build up the back and covings; and if the Fire-place is designed for burning coals, to fix the grate in its proper place, according to the directions already given.—When the width of the Fire-place is reduced, the edges of the covings a A and b B are to make a finish with the front of the jambs.—And in general it will be best, not only for the sake of the appearance of the Chimney, but for other reasons also, to lower the height of the opening of the Fire-place, whenever its width in front is diminished.

Fig. 4. shows a front view of the Chimney after it has been altered according to the directions here given.—By comparing it with fig. 2.

(which shows a front view of the same Chimney before it was altered), the manner in which the opening of the Fire-place in front is diminished may be seen. — In fig. 4. the under part of the door-way by which the Chimney-sweeper gets up the Chimney is represented by white dotted lines. The door-way is represented closed.

I shall finish this chapter with some general observations relative to the subject under consideration; with directions how to proceed where such local circumstances exist as render modifications of the general plan indispensably necessary.

Whether a Chimney be designed for burning wood upon the hearth, or wood, or coals in a grate, the form of the Fire-place is in my opinion, most perfect when THE WIDTH OF THE BACK is equal to the DEPTH OF THE FIRE-PLACE, and the opening of the Fire-place in front equal to THREE TIMES the width of the back, or, which is the same thing, to THREE TIMES THE DEPTH OF THE FIRE-PLACE.

But if the Chimney be designed for burning wood upon the hearth, upon hand irons, or dogs, as they are called, it will sometimes be necessary to accommodate the width of the back to the length of the wood; and when this is the case, the covings must be accommodated to the width of the back, and the opening of the Chimney in front.

When the wall of the Chimney in front, measured from the upper part of the breast of the Chimney to the front of the mantle, is very thin, it may happen, and especially in Chimnies designed for burning wood upon the hearth, or upon dogs, that the depth of the Chimney, determined according to the directions here given, may be too small.

Thus, for example, supposing the wall of the Chimney in front, from the upper part of the breast of the Chimney to the front of the mantle, to be only four inches, (which is sometimes the case, particularly in rooms situated near the top of a house,) in this case, if we take four inches for the width of the throat, this will give eight inches only for the depth of the Fire-place, which would be too little, even were coals to be burnt instead of wood. — In this case I should increase the depth of the Fire-place at the hearth to 12 or 13 inches, and should build the back perpendicular to the height of the top of the burning fuel, (whether it be wood burnt upon the hearth, or coals in a grate,) and then, sloping the back by a gentle inclination forward, bring it to its proper place, that is to say, PERPENDICU-LARLY UNDER THE BACK OF THE THROAT OF THE CHIM-NEY. This slope, (which will bring the back forward four or five inches, or just as much as the depth of the Fire-place is encreased,) though it ought not to be too abrupt, yet it ought to be quite finished at the height of eight or ten inches above the fire, otherwise it may perhaps cause the Chimney to smoke; but when it is very near the fire, the heat of the fire will enable the current of rising smoke to overcome the obstacle which this slope will oppose to its ascent, which it could not do so easily were the slope situated at a greater distance from the burning fuel[2].

Fig. 7, 8, and 9, show a plan, elevation, and section of a Fire-place constructed or altered upon this principal.—The wall of the Chimney in front at a, fig. 9, being only four inches thick, four inches more added to it for the width of the throat would have left the depth of the Fire-place measured upon the hearth b c only eight inches, which would have been too little;—a niche c and e, was therefore made in the new back of the Fire-place for receiving the grate, which niche was six inches deep in the center of it, below 13 inches wide, (or equal in width to the grate,) and 23 inches high; finishing above with a semicirular arch, which, in its highest part, rose seven inches above the upper part of the grate.—The door-way for the Chimney-sweeper, which begins just above the top of the niche, may be seen distinctly in both the figures 8 and 9.—The space marked g, fig. 9, behind this door-way, may either be filled with loose bricks, or may be left void.—The manner in which the piece of stone f, fig. 9, which is put under the mantle of the Chimney to reduce the height of the opening of the Fire-place, is rounded off on the inside in order to give a fair run to the column of smoke in its ascent through the throat of the Chimney, is clearly expressed in this figure.

The plan fig. 7, and elevation fig. 8, show how much the width of the opening of the Fire-place in front is diminished, and how the covings in the new Fire-place are formed.

A perfect idea of the form and dimension of the Fire-place in its original state, as also after its alteration, may be had by careful inspection of these figures.

I have added the drawing fig. 10, merely to show how a fault, which I have found workmen in general whom I have employed in altering Fire-places are very apt to commit, is to be avoided. —In Chimneys like that represented in this figure, where the jambs A and B project far into the room, and where the front edge of the marble slab, o which forms the coving, does not come so far forward as the front of the jambs, the workmen in constructing the new covings are very apt to place them, —not in the line c A, which they ought to do, —but in the line c o, which is a great fault. —The covings of a Chimney should never range BEHIND the front of the jambs, however those jambs may project into the room; —but it is not absolutely necessary that the covings should MAKE A FINISH with the internal front corners of the jambs, or that they should be continues from the back c, quite to the front of the jambs at A. — They may finish in front at a and b, and small corners A, o, a, may be left for placing the shovels, tongs, etc.

Were the new coving to range with the front edge of the old coving o, the obliquity of the new coving would commonly be too great; —or the angle d c o would exceed 135 degrees, WHICH IT NEVER SHOULD DO, —or at least never by more than a very few degrees.

No inconvenience of any importance will arise from making the obliquity of the covings LESS than what is here recommended; but many cannot fail to be produced by making it much greater; — and as I know from experience that workmen are very apt to do this, I have thought it necessary to warn them particularly against it.

Fig. 11. shows how the width and obliquity of the covings of a Chimney are to be accommodated to the width of the back, and to the opening in front and depth of the Fire-place, where the width of the opening of the Fire-place is less than three times the width of the new back. As all those who may be employed in altering Chimneys may not, perhaps, known how to set off an angle of any certain numbers of degrees, —or may not have at hand the instruments necessary for doing it, —I shall here show how an instrument may be made which will be found to be very useful in laying out the work for the bricklayers.

Upon a board about 18 inches wide and four feet long, or upon the floor or a table, draw three equal squares A, B, C, fig. 12. of about 12 or 14 inches each side, placed in a strait line, and touching each other.—From the back corner c of the center square B, draw a diagonal line across the square A, to its outward front corner f, and the adjoining angle formed by the lines d c and c f will be equal to 135 degrees,—the angle which the plane of the back of a Chimney Fire-place ought to make with the plane of its covings.—And a bevel m, n, being made to this angle with thin slips of hard wood, this little instrument will be found to be very useful in marking out on the hearth, with chalk, the plans of the walls which are to form the covings of Fire-places.

As Chimneys which are apt to smoke will require the covings to be placed less obliquely in respect to the back than others which have not that defect, it would be convenient to be provided with several bevels;—three or four, for instance, forming different angles.—That already described, which may be called No. 1. will measure the obliquity of the covings when the Fire-place can be made of the most perfect form:—another No. 2. may be made to a smaller angle, d c e,—and another No. 3. for Chimnies which are very apt to smoke at the still smaller angle d c i.—Or a bevel may be so contrived, by means of a joint, and an arch, properly graduated, as to serve for all the different degrees of obliquity which it may ever be necessary to give to the covings of Fire-places.

Another point of much importance, and particularly in Chimneys which are apt to smoke, is to form the throat of the Chimney properly, by carrying up the back and covings to a proper height. This, workmen are apt to neglect to do, probably on account of the difficulty they find in working where the opening of the canal of the Chimney is so much reduced.—But it is absolutely necessary that these walls should be carried up five or six inches at least above the upper part of the breast of the Chimney, or to that point where the wall which forms the front of the throat begins to rise perpendicularly. —If the workman has intelligence enough to avail himself of the opening which is formed in the back of the Fire-place to give a passage to the Chimney-sweeper, he will find little difficulty in finishing his work in a proper manner.

In placing the plumb-line against the breast of the Chimney, in order to ascertain how far the new back is to be brought forward, great care must be taken to place it at the very top of the breast, where the canal of the Chimney BEGINS TO RISE PERPENDICULARLY; otherwise, when the plumb-line is placed too low, or against the slope of the breast, when the new back comes to be raised to its proper height, the throat of the Chimney will found to be too narrow.

Sometimes, and indeed very often the top of the breast of a Chimney lies very high, or far above the fire (see the figures 13 and 14, where d shows the top of the breast of the Chimney); when this is the case it must be brought lower, otherwise the Chimney will be very apt to smoke.—So much has been said in the First Chapter of this Essay of the advantages to be derived from bringing the throat of a Chimney near to the burning fuel, that I do not think it necessary to enlarge on them in this place,— taking it for granted that the utility and necessity of that arrangement have already been made sufficiently evident;— but a few directions for workmen, to show them how the breast (and consequently the throat) of a Chimney can most readily be lowered, may not be superfluous.

Where the too great height of the breast of a Chimney is owing to the great height of the mantle, (see fig. 13,) or, which is the same thing, of the opening of the Fire-place in front, which will commonly be found to be the case; the only remedy for the evil will be to bring down the mantle lower;—or rather, to make the opening of the Fire-place in front lower, by throwing across the top of this opening, from one jamb to the other, and immediately under the mantle, a very flat arch;—a wall of bricks and mortar, supported on straight bars of iron;—or a piece of stone (h, fig. 13).—When this is done, the slope of the old throat of the Chimney, or of the back side of the mantle, is to be filled up with plaster, so as to form one continued flat, vertical, or upright plane surface with the lower part of the wall of the canal of the Chimney, and a new breast is to be formed lower down, care being taken to round it off properly, and make it finish at the lower surface of the new wall built under the mantle;—which wall forms in fact a new mantle.

The annexed drawing fig. 13, which represents the section of a Chimney in which the breast has been lowered according to the method here described, will show these various alterations in a clear and satisfactory manner. In this figure, as well as in most of the others in this Essay, the old walls are distinguished from the new ones by the manner in which they are shaded;— the old walls being shaded by diagonal lines, and the new ones by vertical lines. The additions, which are formed of plaster, are shaded by dots instead of lines.

Where the too great height of the breast of a Chimney is occasioned, not by the height of the mantle, but by the too great width of the breast, in that case, (which however will seldom be found to occur,) this defect may be remedied by covering the lower part of the breast with a thick coating of plaster, supported, if necessary, by nails or studs, driven into the wall which forms the breast, and properly rounded off at the lower part of the mantle.—See fig. 14.

CHAPTER. III.

Of the cause of the ascent of smoke.
Illustration of the subject by familiar comparisons and
  experiments.
Of chimnies which affect and cause each other to smoke.
Of chimnies which smoke from want of air.
Of the eddies of wind which sometimes blow down chimnies,
  and cause them to smoke.
Explanation of the figures.

Though it was my wish to avoid all abstruse philosophical investigations in this Essay, yet I feel that it is necessary to say a few
words upon a subject generally considered as difficult to be explained, which is too intimately connected with the matter under
consideration to be passed over in silence.— A knowledge of the
cause of the ascent of Smoke being indispensably necessary to those
who engage in the improvement of Fire-places, or who are desirous
of forming just ideas relative to the operations of fire, and the management of heat, I shall devote a few pages to the investigation of
that curious and interesting subject.—And as many of those who
may derive advantage from these inquiries are not much accustomed to philosophical disquisitions, and would not readily com-
prehend either the language or the diagrams commonly used by
scientific writers to explain the phaenomena in question, I shall take
pains to express myself in the most familiar manner, and to use
such comparisons for illustration as may easily be understood.

If small leaden bullets, or large goose shot, be mixed with peas,
and the whole well shaken in a bushel, the shot will separate from
the peas, and will take its place at the bottom of the bushel; forcing
by its greater weight the peas which are lighter, to move upwards,
contrary to their natural tendency, and take their places above.

If water and linseed oil, which is lighter than water, be mixed in a
vessel by shaking them together, upon suffering this mixture to
remain quite, the water will descend and occupy the bottom of the
vessel, and the oil, being forced out of its place by the greater pres-

sure downwards of the heavier liquid, will be obliged to rise and swim on the surface of the water.

If a bottle containing linseed oil be plunged in water with its mouth upwards, and open, the oil will ascent out of the bottle, and passing upwards through the mass of water, in a continued stream, will spread itself over its surface.

In like manner when two fluids of any kind, of different densities, come into contact, or are mixed with each other, that which is the lightest will be forced upwards by that which is the heaviest.

And as heat rarefies all bodies, fluids as well as solids, air as well as water, or mercury,—it follows that two portions of the same fluid, at different temperatures, being brought into contact with each other, that portion which is the hottest being more rarefied or specifically LIGHTER than that which is colder, must be forced upwards by this last.—And this is what always happens in fact.

When hot water and cold water are mixed, the hottest part of the mixture will be found to be at the surface above;—and when cold air is admitted into a warmed room, it will always be found to take its place at the bottom of the room, the warmer air being in part expelled, and in part forced upwards to the top of the room.

Both air and water being transparent and colourless fluids, their internal motions are not easily discovered by the sight, and when these motions are very slow, they make no impression whatever on any of our senses, consequently they cannot be detected by us without the aid of some mechanical contrivance:— But where we have reason to think that those motions exist, means should be sought, and may often be found, for rendering them perceptible.

If a bottle containing hot water tinged with log-wood, or any other colouring drug, be immersed, with its mouth open, and upwards, into a deep glass jar filled with cold water, the ascent of the hot water from the bottle through the mass of cold water will be perfectly visible through the glass.— Now nothing can be more evident than that both of these fluids are forced, or PUSHED, and not DRAWN upwards.—Smoke is frequently said to be drawn up the Chimney;—and that a Chimney draws well, or ill;—but these are careless expressions, and lead to very erroneous ideas respecting the cause of the ascent of Smoke; and consequently tend to prevent the progress of improvements in the management of fires.—The experiment just mentioned with the coloured water is very striking and beautiful, and it is well calculated to give a just idea of the cause of the ascent of Smoke. The cold water in the jar, which, in consequence of its superior weight or density, forces the heated and rarefied water in the bottle to give place to it, and to move upwards out of its way, may represent the cold air of the atmosphere, while the rising column of coloured water will represent the column of Smoke which ascends from a fire.

If Smoke required a Chimney to DRAW it upwards, how happens it that Smoke rises from a fire which is made in the open air, where there is no Chimney?

If a tube, open at both ends, and of such a length that its upper end be below the surface of the cold water in the jar, be held vertically over the mouth of the bottle which contains the hot coloured water, the hot water will rise up through it, just a smoke rises in a Chimney.

If the tube be previously heated before it is plunged into the cold water, the ascent of the hot coloured water will be facilitated and accelerated, in like manner as Smoke is known to rise with greater facility in a Chimney which is hot, than in one in which no fire has been made for a long time.—But in neither of these cases can it, with any propriety, be said, that the hot water is DRAWN up the tube.— The hotter the water in the bottle is, and the colder that in the jar, the greater will be the velocity with which the hot water will be forced up through the tube; and the same holds of the ascent of hot Smoke in a Chimney.—When the fire is intense, and the weather very cold, the ascent of the Smoke is very rapid; and under such circumstances Chimneys seldom smoke.

As the cold water of the jar immediately surrounding the bottle which contains the hot water, will be heated by the bottle while the other parts of the water in the jar will remain cold, this water so heated, becoming specifically lighter than that which surrounds it, will be forced upwards; and if it finds its way into the tube will rise up through it with the coloured hot water.—The warmed air of a room heated by an open Chimney Fire-place has always a tendency to rise, (if I may use that inaccurate expression,) and finding its way into the Chimney frequently goes off with the Smoke.

What has been said, will, I flatter myself, be sufficient to explain and illustrate, in a clear and satisfactory manner, the cause of the ascent of Smoke; and just ideas upon that subject are absolutely necessary in order to judge, with certainty, of the merit of any scheme proposed for the improvement of Fire-places; or to take effectual measures, in all cases, for curing smoking Chimnies. — For though the perpetual changes and alterations which are produced by accident, whim, and caprice, do sometimes lead to useful discoveries, yet the progress of improvement under such guidance must be exceedingly slow, fluctuating, and uncertain.

As to the causes of the smoking of Chimnies, they are very numerous, and various; but as a general idea of them may be acquired from what has already been said upon that subject in various parts of this Essay, and as they may, in all cases, (a very few only excepted,) be completely remedied by making the alterations in Fire-places here pointed out; I do not think it necessary to enumerate them all in this place, or to enter into those long details and investigations which would be required to show the precise manner in which each of them operates, either alone, or in conjunction with others.

There is however one cause of smoking Chimnies which I think it is necessary to mention more particularly. — In modern built houses, where the doors and windows are generally made to close with such accuracy that no crevice is left for the passage of the air from without, the Chimnies in rooms adjoining to each other, or connected by close passages, are frequently found to affect each other, and this is easy to be accounted for. — When there is a fire burning in one of the Chimnies, as the air necessary to supply the current up the Chimney where the fire burns cannot be had in sufficient quantities from without, through the very small crevices of the doors and windows, the air in the room becomes rarefied, not by heat, but by subtraction of that portion of air which is employed in keeping up the fire, or supporting the combustion of the fuel, and in consequence of this rarefaction, its elasticity is diminished, and being at last overcome by the pressure of the external air of the atmosphere, this external air rushes into the room by the only passage left for it, namely, by the open Chimney of the neighbouring room:—And the flow of air into the Fire-place, and up the Chimney where the fire is burning being constant, this expence of air is supplied by a continued current down the other Chimney.

If an attempt be made to light fires in both Chimnies at the same time, it will be found to be very difficult to get the fires to burn, and the rooms will both be filled with Smoke.

One of the fires,—that which is made in the Chimney where the construction of the Fire-place is best adapted to facilitate the ascent of the Smoke,—or if both Fire-places are on the same construction,—that which has the wind most favourable, or in which the fire happens to be soonest kindled,—will overcome the other, and cause its Smoke to be beat back into the room by the cold air which descends through the Chimney.—The most obvious remedy in this case is to provide for the supply of fresh air necessary for keeping up the fires by opening a passage for the external air into the room by a shorter road than down one of the Chimnies; and when this is done, both Chimnies will be found to be effectually cured.

But Chimnies so circumstanced may very frequently be prevented from smoking even without opening any new passage for the external air, merely by diminishing the draught, (as it is called,) up the Chimnies; which can best be done by altering both Fire-places upon the principles recommended and fully explained in the foregoing Chapters of this Essay.

Should the doors and windows of a room be closed with so much nicety as to leave no crevices by which a supply of air can enter sufficient for maintaining the fire, AFTER THE CURRENT OF AIR UP THE CHIMNEY HAS BEEN DIMINISHED AS MUCH AS POSSIBLE BY DIMINISHING THE THROAT OF THE FIRE-PLACE; in that case there would be no other way of preventing the Chimney from smoking but by opening a passage for the admission of fresh air from without;—but this, I believe, will very seldom be found to be the case.

A case more frequently to be met with is where currents of air set down Chimnies in consequence of a diminution and rarefaction of the air in a room, occasioned by the doors of the room opening into passages or courts where the air is rarefied by the action of some particular winds. In such cases the evil may be remedied, either by causing the doors in question to close more accurately,—or, (which will be still more effectual,) by giving a supply of air to the passage or court which wants it, by some other way.

Where the top of a Chimney is commanded by high buildings, by clifts, or by high grounds, it will frequently happen, in windy weather, that the eddies formed in the atmosphere by these obstacles will blow down the Chimney, and beat down the smoke into the room.—This it is true will be much less likely to happen when the throat of the Chimney is contracted and properly formed than when it is left quite open, and the Fire-place badly constructed; but as it is POSSIBLE that a Chimney may be so much exposed to these eddies in very high winds as to be made to smoke sometimes when the wind blows with violence from a certain quarter, it is necessary to show how the effects of those eddies may be prevented.

Various mechanical contrivances have been imagined for preventing the wind from blowing down Chimnies, and many of them have been found to be useful;—there are, however, many of these inventions, which, though they prevent the wind from blowing down the Chimney, are so ill-contrived on other accounts as to obstruct the ascent of the Smoke, and do more harm than good.

Of this description are all those Chimney-pots with flat horizontal plates or roofs placed upon supporters just above the opening of the pot;—and most of the caps which turn with the wind are not much better.—One of the most simple contrivances that can be made use of, and which in most cases will be found to answer the purpose intended as well or better than more complicated machinery, is to cover the top of the Chimney with a hollow truncated pyramid or cone, the diameter of which above, or opening for the passage of the Smoke, is about 10 or 11 inches. —This pyramid, or cone, (for either will answer,)—should be of earthen ware, or of cast iron;—its perpendicular height may be equal to the diameter of its opening above, and the diameter of its opening below equal to three times its height.—It should be placed upon the top of the Chimney, and it may be contrived so as to make a handsome finish to the brickwork.—Where several flews come out near each other, or in the same stack of Chimnies, the form of a pyramid will be better than that of a cone for these covers.

The intention of this contrivance is, that the winds and eddies which strike against the oblique surface of these covers may be reflected upwards instead of blowing down the Chimney. — The invention is by no means new, but it has not hitherto been often put in practice. — As often as I have seen it tried it has been found to be of use; I cannot say, however, that I was ever obliged to have recourse to it, or to any similar contrivance; and if I forbear to enlarge upon the subject of these inventions, it is because I am persuaded that when Chimnies are properly constructed IN THE NEIGHBOURHOOD OF THE FIRE-PLACE little more will be necessary to be done at the top of the Chimney than to leave it open.

I cannot conclude this Essay without again recommending, in the strongest manner, a careful attention to the management of fires in open Chimnies; for not only the quantity of heat produced on the combustion of fuel depends much on the manner in which the fire is managed, but even of the heat actually generated a very small part only will be saved, or usefully employed, when the fire is made in a careless and slovenly manner.

In lighting a coal fire more wood should be employed than is commonly used, and fewer coals; and as soon as the fire burns bright, and the coals are well lighted, and NOT BEFORE, more coals should be added to increase the fire to its proper size[3].

The enormous waste of fuel in London may be estimated by the vast dark cloud which continually hangs over this great metropolis, and frequently overshadows the whole country, far and wide; for this dense cloud is certainly composed almost entirely of UNCONSUMED COAL, which having stolen wings from the innumerable fires of this great city has escaped by the Chimnies, and continues to sail about in the air, till having lost the heat which gave it volatility, it falls in a dry shower of extremely fine black dust to the ground, obscuring the atmosphere in its descent, and frequently changing the brightest day into more than Egyptian darkness.

I never view from a distance, as I come into town, this black cloud which hangs over London, without wishing to be able to compute the immense number of chaldrons of coals of which it is composed;

for could this be ascertained, I am persuaded so striking a fact would awaken the curiosity, and excite the astonishment of all ranks of the inhabitants; and PERHAPS turn their minds to an object of economy to which they have hitherto paid little attention.

Conclusion.

Though the saving of fuel which will result from the improvements in the forms of CHIMNEY FIRE-PLACES here recommended will be very considerable, yet I hope to be able to show in a future Essay, that still greater savings may be made, and more important advantages derived from the introduction of improvements I shall propose in KITCHEN FIRE-PLACES.

I hope likewise to be able to show in an Essay on COTTAGE FIRE-PLACES, which I am now preparing for publication, that THREE QUARTERS, at least, of the fuel which cottagers now consume in cooking their victuals, and in warming their dwellings, may with great ease, and without any expensive apparatus, be saved.

END OF THE FOURTH ESSAY.

EXPLANATION OF THE FIGURES

[IMAGE]

Fig. 1.
The plan of a Fire-place on the common construction.
A B, the opening of the Fire-place in front.
C D, the back of the Fire-place.
A C and B D, the covings.
See page 341.

[IMAGE]

Fig. 2. This figure shows the elevation, or front view of a Fire-place on the common construction. See page 341.

[IMAGE]

Fig. 3. This Figure shows how the Fire-place represented by the Fig. 1, is to be altered in order to its being improved.

A B is the opening in front, — C D, the back, and A C and B D, the covings of the Fire-place in its original state.

a b, its opening in front, — i k, its back, — and a i and b k, its covings after it has been altered, e is a point upon the hearth upon which a plum suspended from the middle of the upper part of the breast of the Chimney falls. The situation for the new back is ascertained by taking the line e f equal to four inches. The new back and covings are represented as being built of bricks; — and the space between these and the old back and covings as being filled up with rubbish. See page 342.

[IMAGE]

Fig. 4. This Figure represents the elevation or front view of the Fire-place Fig. 3. after it has been altered. The lower part of the door-way left for the Chimney-sweeper is shown in this Figure by white dotted lines. See page 344.

[IMAGE]

Fig. 5. This Figure shows the section of a Chimney Fire-place and of a part of the canal of the Chimney, on the common construction.

a b is the opening in front; b c, the depth of the Fire-place at the hearth; d, the breast of the Chimney.

d e, the throat of the Chimney, and d f, g e, a part of the open canal of the Chimney.

[IMAGE]

Fig. 6.
Shows a section of the same Chimney after it has been altered.

k l is the new back of the Fire-place; l i, the tile or stone which closes the door-way for the Chimney-sweeper; d i, the throat of the Chimney, narrow to four inches; a, the mantle, and h, the new wall made under the mantle to diminish the height of the opening of the Fire-place in front.

N.B. These two Figures are sections of the same Chimney which is represented in each of the four preceding Figures.

[IMAGE]

Fig. 7.

This Figure represents the ground plan of a Chimney Fire-place in which the grate is placed in a niche, and in which the original width A B of the Fire-place is considerably diminished.

a b is the opening of the Fire-place in front after it has been altered, and d is the back of the niche in which the grate is placed. See page 347.

[IMAGE]

Fig. 8.
Shows a front view of the same Fire-place after it has been altered; where may be seen the grate, and the door-way for the Chimney-sweeper. See page 347.

[IMAGE]

Fig. 9. Shows a section of the same Fire-place, c d e being a section of the niche, g the door-way for the Chimney-sweeper, closed by a piece of the fire-stone, and f the new wall under the mantle by which the height of the opening of the Fire-place in front is diminished. See page 347.

[IMAGE]

Fig. 10. This Figure shows how the covings are to be placed when the front of the covings (a and b) do not come so far forward as the front of the opening of the Fire-place, or the jambs (A and B). See page 348.

[IMAGE]

Fig. 11. This Figure shows how the width and obliquity of the covings are to be accommodated to the width of the back of a Fire-place, in cases where it is necessary to make the back very wide. See page 349.

[IMAGE]

Fig. 12. This Figure shows how an instrument called a bevel (m n), useful in laying out the work, in altering Chimney Fire-places, may be constructed. See page 349.

[IMAGE]

Fig. 13. This shows how, when the breast of a Chimney (d) is too high, it may be brought down by means of a wall (h) placed under the mantle, and a coating of plaster, which in this Figure is represented by the part marked by dots. See page 351.

[IMAGE]

Fig. 14. This shows how the breast of a Chimney may be brought down merely by a coating of plaster. See page 351.

Footnotes for essay IV.

[1] Eves and Sutton, bricklayers, Broad Sanctuary, Westminster, have alone altered above 90 Chimnies. — The experiment was first made in London at Lord Palmerston's house in Hanover-square; — then two Chimnies were altered in the house of Sir John Sinclair, Baronet, President of the Board of Agriculture; one in the room in which the Board meets, and the other in the Secretary's room; which last being much frequented by persons from all parts of Great Britain, it was hoped that circumstances would tend much to expedite the introduction of these improvements in various parts of the kingdom. Several Chimnies were altered in the house of Sir Joseph Banks, Baronet, K. B. President of the Royal Society. Afterwards a number were altered in Devonshire-house; — in the house of Earl Besborough, in Cavendish-square, and at his seat at Roehampton; — at Holywell-house, near St. Alban's, the seat of the Countess Dowager Spencer: — at Melbourne-house; — at Lady Templeton's in Portland-place; — at Mrs Montagu's in Portman-square; — at Lord Sudley's, in Dover-street: — at the Marquis of Salisbury's seat at Hatfield, and at his house in town; — at Lord Palmerston's seat at Broadlands, near Southampton, and at several gentlemen's houses in that neighbourhood; — and a great many others; but it would be tiresome to enumerate them all; and even these are mentioned merely for the satisfaction of those who may wish to make inquiries respecting the success of the experiments.

[2] Having been obliged to carry backward the Fire-place in the manner here described, in order to accommodate it to a Chimney whose walls in front were remarkably thin, — I was surprised to find upon lighting the fire that it appeared to give out more heat into the room than any Fire-place I had ever constructed. — This effect was quite unexpected; but the cause of it was too obvious not to be immediately discovered. — The flame rising from the fire broke against the part of the back which sloped forward over the fire, and this part of the back being soon very much heated, and in consequence of its being very hot, (and when the fire burnt bright it was frequently quite red hot,) it threw off into the room a great deal of radiant heat. — It is not possible that this oblique surface (the slope of the back of the Fire-place) could have been heated red-hot MERELY by the radiant heat projected by the burning fuel, for other parts of the Fire-place nearer the fire, and better situated for receiv-

ing radiant heat, were never found to be so much heated;—and hence it appears that the combined heat in the current of smoke and hot vapour which rises from an open fire MAY BE, at least IN PART, stopped in its passage up the Chimney, changing into radiant heat, and afterwards thrown into the room.—This opens up a new and very interesting field for experiment, and bids fair to lead to important improvements in the construction of Fire-places.—I have of late been much engaged in these investigations, and am now actually employed daily in making a variety of experiments with grates and Fire-places, upon different constructions, in the room I inhabit in the Royal Hotel in Pall Mall;—and Mr. Hopkins of Greek-street Soho, Ironmonger to his Majesty, and Mrs. Hempel, at her Pottery at Chelsea, are both at work in their different lines of business, under my direction, in the construction of Fire-places upon a principle entirely new, and which, I flatter myself, will be found to be not only elegant and convenient, but very economical.—But as I mean soon to publish a particular account of these Fire-places,—with drawings and ample directions for constructing them, I shall not enlarge farther on the subject in this place.—It may however not be amiss just to mention here, that these new-invented Fire-places not being fixed to the walls of the Chimney, but merely set down upon the hearth, may be used in any open Chimney: and that Chimnies altered or constructed on the principles here recommended are particularly well adapted for receiving them.

The Public in general, and more particularly those Tradesmen and Manufacturers whom it may concern, are requested to observe, that as the Author does not intent to take out himself, or to suffer others to take out, any patent for any invention of his which may be of public utility, all persons are at full liberty to imitate them, and vend them, for their own emolument, when and where, and in any way they may think proper; and those who may wish for any further information respecting any of those inventions or improvements will receive (gratis) all the information they can require by applying to the Author, who will take pleasure in giving them every assistance in his power.

[3] Kindling balls composed of equal parts of coal,—charcoal, — and clay, the two former reduced to a fine powder, well mixed and kneaded together with the clay moistened with water, and then

formed into balls of the size of hens eggs, and thoroughly dried, might be used with great advantage instead of wood for kindling fires. These kindling balls may be made so inflammable as to take fire in an instant and with the smallest spark, by dipping them in a strong solution of nitre and then drying them again, and they would neither be expensive nor liable to be spoiled by long keeping. Perhaps a quantity of pure charcoal reduced to a very fine powder and mixed with the solution of nitre in which they are dipped would render them still more inflammable.

I have often wondered that no attempts should have been made to improve the fires which are made in the open Chimnies of elegant apartments, by preparing the fuel; for nothing surely was ever more dirty, inelegant, and disgusting than a common coal fire.

Fire balls of the size of goose eggs, composed of coal and charcoal in powder, mixed up with a due proportion of wet clay, and well dried, would make a much more cleanly, and in all respects a pleasanter fire than can be made with crude coals; and I believe would not be more expensive fuel. In Flanders and in several parts of Germany, and particular in the Dutchies of Juliers and Bergen, where coals are used as fuel, the coals are always prepared before they are used, by pounding them to a powder, and mixing them up with an equal weight of clay, and sufficient quantity of water to form the whole into a mass which is kneaded together and formed into cakes; which cakes are afterwards well dried and kept in a dry place for use. And it has been found by long experience that the expense attending this preparation is amply repaid by the improvement of the fuel. The coals, thus mixed with the clay, not only burn longer, but give much more heat than when they are burnt in their crude state.

It will doubtless appear extraordinary to those who have not considered the subject with some attention, that the quantity of heat produced in the combustion of any quantity of coals should be increased by mixing the coals with clay, which is certainly an incombustible body;—but the phenomenon may, I think, be explained in a satisfactory manner.

The heat generated in the combustion of any small particle of coal existing under two distinct forms, namely, in that which is COM-

BINED with the flame and smoke which rise from the fire, and which if means are not found to stop it, goes off immediately by the Chimney and is lost,—and the RADIANT HEAT which is sent off from the fire, in all directions in right lines:—I think it reasonable to conclude, that the particles of clay which are surrounded on all sides by the flame arrest a part at least of the combined heat, and prevent its escape; and this combined heat, so arrested, heating the clay red hot, is retained in it, and being changed by this operation to radiant heat, is afterwards emitted, and may be directed, and employed to useful purposes.

In composing fire balls, I think it probable that a certain proportion of chaff—of straw cut very fine, or even saw dust, might be employed with great advantage. I wish those who have leisure would turn their thoughts to this subject, for I am persuaded that very important improvements would result from a thorough investigation of it.

CONTENTS of ESSAY V.

A SHORT ACCOUNT of SEVERAL PUBLIC INSTITUTIONS
Lately formed in Bavaria.
together with the
APPENDIX to the FIRST VOLUME.

Account I
 A Short Account of the military academy at munich

Account II
 An account of the means used to improve the bread of horses,
  and horned cattle, in Bavaria and the Palatinate.

Account III
 An account of the measures adopted for putting an end to usury at
  Munich.

Account IV
 An account of a scheme for employing the soldiery in Bavaria in

repairing the highways and public roads.

APPENDIX

No. I

Address and petition to all inhabitants and citizens of Munich,
in the name of the real poor and distressed.

No. II

Subscription lists distributed among the inhabitants of Munich,
in the month of January 1790, when the establishment for the
relief of the poor in that city was formed.

No. III

An account of the receipts and expenditures of the institution
for the poor at Munich during five years.

No. IV

Certificate relative to the expence of fuel in the public kitchen
of the military workhouse at Munich.

No. V

Printed form for the descriptions of the poor.

No. VI

Printed form for spin-tickets, such as are used at the military
workhouse at Munich.

No. VII

An Account of experiments made at the bakehouse of the military
workhouse at Munich, November the 4th and 5th, 1794.

No. VIII

Account of the persons in the house of industry in Dublin the
30th of April 1796, and of the details of the manner and expence
of feeding them.

No. IX

An account of an experiment made (under the direction of the
author,) in the kitchen of the house of industry at Dublin,
in cooking for the poor.

ESSAY V.

A short Account of the MILITARY ACADEMY at MUNICH.

Though it is certain that too much learning is rather disadvantageous than otherwise to the lower classes of the people;—that the introduction of a spirit of philosophical investigation,—literary amusement,—and metaphysical speculation among those who are destined by fortune to gain their livelihood by the sweat of their brow, rather tends to make them discontented and unhappy, than to contribute any thing to their real comfort and enjoyments; yet there appears, now and then, a native genius in the most humble stations, which it would be a pity not to be able to call forth into activity. It was principally with a view to bring forward such extraordinary talents, and to employ them usefully in the public service, that the Military Academy at Munich was instituted. This Academy, which consists of 180 eleves or pupils, is divided into three classes. The first class, which is designed for the education of orphans and other children of the poorer class of Military Officers, and those employed in the Civil Departments of the State, consists of thirty pupils, who are received gratis, from the age of eleven to thirteen years, and who remain in the Academy for years. The second class, which is designed to assist the poorer nobility, and less opulent among the merchants, citizens, and servants of government, in giving their sons a good general education, consists of sixty pupils, who are received from the age of eleven to fifteen years, and who pay to the Academy twelve florins a month; for which sum they are fed, clothed, and instructed. The third class, consisting of ninety pupils, from the age of fifteen to twenty years, who are all admitted gratis, is designed to bring forward such youths among the lower classes of the people as show evident signs of UNCOMMON TALENTS and genius, joined to a sound constitution of body, and a good moral character.

All Commanding Officers of regiments, and Public Officers in Civil departments, and all Civil Magistrates, are authorised and INVITED to recommend subjects for this class of the Academy, and they are not confined in their choice to any particular ranks of society, but they are allowed to recommend persons of the lowest extraction, and most obscure origin. Private soldiers, and the children of soldiers, and even the children of the meanest mechanics and day-labourers, are admissible, provided they possess the necessary requisites; namely, VERY EXTRAORDINARY NATURAL GENIUS, a healthy constitution, and a good character; but if the subject recommended should be found wanting in any of these requisite qualifications, he would not only be refused admittance into the Academy, but the person who recommended him would be very severely reprimanded.

The greatest severity is necessary upon these occasions, otherwise it would be impossible to prevent abuses. An establishment, designed for the encouragement of genius, and for calling forth into public utility talents which would otherwise remain buried and lost in obscurity, would soon become a job for providing for relations and dependants.

One circumstance, relative to the internal arrangement of this Academy, may, perhaps, be though not unworthy of being particularly mentioned, and that is the very moderate expence at which the institution is maintained. By a calculation, founded upon the experience of four years, I find that the whole Academy, consisting of 180 pupils, with professors and masters of every kind, servants, clothing, board, lodging, fire-wood, light, repairs, and every other article, house-rent alone excepted, amounts to no more than 28,000 florins a-year, which is no more than 155 florins, or about fourteen pounds sterling a-year for each pupil; a small sum indeed, considering the manner in which they are kept, and the education they receive.

Though this Academy is called a Military Academy, it is by no means confined to the education of those who are destined for the army; but it is rather an establishment of general education, where the youth are instructed in every science, and taught every bodily exercise, and personal accomplishment, which constitute a liberal education; and which fits them equally for the station of a private gentleman,—for the study of any of the learned professions,—or for any employment, civil or military, under the government.

As this institution is principally designed as a nursery for genius,—as a gymnasium for the formation of men,—for the formation of REAL MEN, possessed of strength and character, as well as talents and accomplishments, and capable of rendering essential service to the state; at all public examinations of the pupils, the heads of all the pupil departments are invited to be present, in order to witness the progress of the pupils, and to mark those who discover talents peculiarly useful in any particular departments or public employment.

How far the influence of this establishment may extend, time must discover. It has existed only six years; but even in that short period, we have had several instances of very uncommon talents having been called forth into public view, from the most obscure situations. I only wish that the institution may be allowed to subsist.

An Account of the Means used to improve the BREED of HORSE, and HORNED CATTLE, in BAVARIA and the PALATINATE.

Through many parts of the Elector's dominions are well adapted for the breeding of fine horses, and great numbers of horses are actually bred[1]; yet no great attention had for many years been paid to the improvement of the breed; and most of the horses of distinction, such as were used by the nobility as saddle-horses and coach-horses, were imported from Holstein and Mecklenburg.

Being engaged in the arrangement of a new military system for the country, it occurred to me that, in providing horses for the use of the army, and particularly for the train of artillery, such measures might be adopted as would tend much to improve the breed of horses throughout the country; and my proposals meeting with the approbation of his Most Serene Electoral Highness, the plan was carried into execution in the following manner:

A number of fine mares were purchased with money take from the military chest, and being marked with an M (the initial of

Militaria), in a circle, upon the left hip, with a hot iron, they were given to such of the peasants, owning or leasing farms proper for breeding good horses, as applied for them. The conditions upon which these brood mares were given away were as follows:

They were, in the first place, given away gratis, and the person who received one of these mares is allowed to consider her as his own property, and use her in any kind of work he thinks proper; he is, however, obliged not only to keep her, and not to sell her, or give her away, but he is also under obligations to keep her as a brood mare, and to have her regularly covered every season, by a stallion pointed out to him by the commissioners, who are put at the head of this establishment. If she dies, he must replace her with another brood mare, which must be approved by the commissioners, and then marked. — If one of these mares should be found not to bring good colts, or to have any blemish, or essential fault or imperfection, she may be changed for another.

The stallions which are provided for these mares, and which are under the care of the commissioners, are provided gratis; and the foals are the sole property of those who keep the mares, and they may sell them, or dispose of them, when and where, and in any way they may think proper, in the same manner as they dispose of any other foal, brought by any other mare.

In case the army should be obliged to take the field, AND IN NO OTHER CASE WHATEVER, those who are in possession of these mares are obliged either to return them, or to furnish, for the use of the army, another horse fit for the service of the artillery.

The advantages of this arrangement to the army are obvious. In the case of an emergency, horses are always at hand, and these horses being bought in time of peace cost much less than it would be necessary to pay for them, were they to be purchased in a hurry upon the breaking out of a war, upon which occasions they are always dear, and sometimes not to be had for money.

It may perhaps be objected, that the money being laid out so long before the horses are wanted, the loss of the interest of the purchase-money ought to be taken into account; but as large sums of money must always be kept in readiness in the military chest, to enable the army to take the field suddenly, in case it should be necessary; and as a part of this money must be employed in the purchase of horses; it may as well be laid out beforehand, as to lie dead in the military chest till the horses are actually wanted; consequently the objection is not founded.

I wish I could say, that this measure had been completely successful; but I am obliged to own, that it has not answered my expectations. Six hundred mares only were at first ordered to be purchased and distributed; but I had hopes of seeing that number augmented soon to as many thousands; and I had even flattered myself with an idea of the possibility of placing in this manner among the peasants, and consequently having constantly in readiness, without any expence, a sufficient number of horses for the whole army; for the cavalry as well as for the artillery and baggage; and I had formed a plan for collecting together and exercising, every year, such of these horses as were destined for the service of the cavalry, and for permitting their riders to go on furlough with their horses: in short, my views went to the forming of an arrangement, very economical, and in many respects similar to that of the ancient feudal military system; but the obstinacy of the peasantry prevented these measures being carried into execution. Very few of them could be prevailed upon to accept of these horses; and in proportion as the terms upon which they were offered to them were apparently advantageous, their suspicions were increased, and they never would be persuaded that there was not some trick at the bottom of the scheme to over-reach them.

It is possible that their suspicions were not a little increased by the malicious insinuations of persons, who, from motives too obvious to require any explanation, took great pains at that time to render abortive every public undertaking in which I was engaged. But be that as it may, the fact is, I could never find means to remove these suspicions entirely, and I met with so much difficulty in carrying the measure into execution, that I was induced at last to abandon it, or rather to postpone its execution to a more favourable moment. Some few mares (two or three hundred) were placed in different parts of the country; and some very fine colts have been produced from them, during the six years that have elapsed since this institution was formed; but these slow advances do not satisfy the ardour of my zeal for improvement; and if means are not found to accelerate them, Bavaria, with all her natural advantages for breeding fine horses, must be obliged, for many years to come, to continue to import horses from foreign countries.

My attempts to improve the breed of horned cattle, though infinitely more confined, have been proportionally much more successful. Upon forming the public garden at Munich, as the extent of the grounds is very considerable, the garden being above six English miles in circumference, and the soil being remarkably good, I had an opportunity of making, within the garden, a very fine and a very valuable farm; and this farm being stocked with about thirty of the finest cows that could be procured from Switzerland, Flanders, Tyrol, and other places upon the Continent famous for a good breed of horned cattle; and this flock being refreshed annually with new importations of cows as well as bulls, all the cows which are produced, are distributed in the country, being sold to any person of the country who applies for them, AND WITH PROMISE TO REAR THEM, at the same low prices at which the most ordinary calves of the common breed of the country are sold to the butchers.

Though this establishment has existed only about six years, it is quite surprising what a change it has produced in the country. As there is a great resort to Munich from all parts of the country, it being the capital, and the residence of the Sovereign, the new English garden (as it is called), which begins upon the ramparts of the town, and extends near two English miles in length, and is always kept open, is much frequented, and there are few who go into the garden without paying a visit to the cows, which are always at home. Their stables, which are concealed in a thick wood behind a public coffee-house or tavern in the middle of the garden, are elegantly fitted up and kept with great care; and the cows, which are not only large, and remarkably beautiful, but are always kept perfectly clean, and in the highest condition, are an object of public curiosity. Those who are not particularly interested in the improvement of cattle, go to see them as beautiful and extraordinary animals; but farmers and connoisseurs go to EXAMINE them,—to compare them with each other,—and with the common breed of the country, and to get information with respect to the manner of feeding them, and the profits derived from them; and so rapidly has the flame of improvement spread throughout every part of Bavaria from this small spark, that I have no doubt but in a very few years the breed of horned cattle will be quite changed.

Not satisfied with the scanty supply furnished from the farm in the English garden, several of the nobility, and some of the most wealthy and enterprising of the farmers, are sending to Switzerland, and other distant countries famous for fine cattle, for cows and bulls; and the good effects of these exertions are already visible in many parts of the country.

How very easy would it be by similar means to introduce a spirit of improvement in any country! and where sovereigns do not make public gardens to bring together a concourse of people, individuals might do it by private subscription, or at least they might unite together and rent a large farm in the neighbourhood of the capital, for the purpose of making useful experiments. If such a farm were well managed, the produce of it would be more than sufficient to pay all the expenses attending it; and if the grounds and fields were laid out with taste—if good roads for carriages and for those who ride on horseback were made round it, and between all the fields—if the stables were elegantly fitted up—filled with beautiful cattle, kept perfectly clean and neat; and if a handsome inn were erected near the buildings of the farm, where those who visited it might be furnished with refreshment, it would soon become a place of public resort and improvements in agriculture would become A FASHIONABLE AMUSEMENT; the ladies even would take pleasure in viewing from their carriages the busy and most interesting scenes of rural industry, and it would no longer be thought vulgar to understand the mysteries of Ceres.

Why should not Parliament purchase, or rent such a farm in the neighbourhood of London, and put it under the direction of the Board of Agriculture? The expence would be but a mere trifle, if any thing, and the institution would not only be useful, but extremely interesting; and it would be an inexhaustible source of rational and innocent amusement, as well as of improvement to vast numbers of the most respectable inhabitants of this great metropolis.

In former times, statesmen considered the amusement of the public as an object of considerable importance, and pains were taken to render the public amusements useful in forming the national character.

An Account of the Measures adopted for putting an End to USURY at MUNICH.

Another measure, more limited in its operations than those before mentioned, but which notwithstanding was productive of much good, was adopted, in which a part of the treasure which was lying dead in the military chest was usefully employed for the relief of a considerable number of individuals, employed in subordinate stations under the government, who stood in great need of assistance.

A practice productive of much harm to the public service, as well as to individuals, had prevailed for many years in Bavaria in almost all the public departments of the state, that of appointing a great number of supernumerary clerks, secretaries, counsellors, etc. who, serving without pay, or with only small allowances, were obliged, in order to subsist till such time as they should come into the receipt of the regulated salaries annexed to their offices, to contract debts to a considerable amount; and as many of them had no other security to give for the sums borrowed, than their promise to repay them when it should be in their power, no money-lender who contented himself with legal interest for his money would trust them; and of course they were obliged to have recourse to Jews and other usurers, who did not afford them the temporary assistance they required, but upon the most exorbitant and ruinous conditions; so that these unfortunate people, instead of finding themselves at their ease upon coming into possession of the emoluments of their offices, were frequently so embarrassed in their circumstances as to be obliged to mortgage their salaries for many months to come, to raise money to satisfy their clamorous creditors; and from this circumstance, and from the general prevalence of luxury and dissipation among all ranks of society, the anticipation of salaries had become so prevalent, and the conditions upon which money was advanced upon such security was so exorbitant, that this alarming evil called for the most serious attention of the government.

The interest commonly paid for money, advanced upon receipts for salaries, was 5 PER CENT. PER MONTH, or three creutzers, for the florin; and there were instances of even much larger interest being given.

The severest laws had been made to prevent these abuses, but means were constantly found to evade them; and, instead of putting an end to the evil, they frequently served rather to increase it.

It occurred to me, that as any tradesman may be ruined by another who can afford to undersell him, so it might be possible to ruin the usurers, by setting up the business in opposition to them, and furnishing money to borrowers upon more reasonable terms. In order to make this experiment, a caise of advance (Vorschuss Cassa), containing 30,000 florins, was established at the military pay-office, where any person in the actual receipt of a salary or pension under government, in any department of the state, civil or military, might receive in advance, upon his personal application, his salary or pension for one or for two months upon a deduction of interest at the rate of 5 PER CENT. PER ANNUM, or one twelfth part of the interest commonly extorted by the Jews and other usurers upon those occasions.

The great number of persons who have availed themselves of the advantages held out to them by this establishment, and who still continue to avail themselves of them, shows how effectually the establishment has been to remedy the evil it was designed to eradicate.

The number of persons who apply to this chest for assistance each month, is at a medium from 300 to 400, and the sums actually in advance, amount in general to above 20,000 florins.

As no money is advanced from this chest but upon government securities that is to say, upon receipts for salaries, and pensions, there is no risque attending the operation; and as the interest arising from the money advanced, is more than sufficient to defray the expence of carrying on the business, there is no loss whatever attending it.

An Account of a SCHEME for employing the SOLDIERY in BAVARIA in repairing the Highways and Public Roads.

I had formed a plan, which, if it had been executed, would have rendered the military posts or patroles of cavalry established in all parts of the Elector's dominions much more interesting, and more useful[2]. I wished to have employed the soldiery exclusively in the repairs of all the highways in the country, and to have united this undertaking with the establishment of permanent military stations, on all the high roads, for the preservation of order and public tranquillity.

It is a great hardship upon the inhabitants in any country to be obliged to leave their own domestic affairs, and turn out with their cattle and servants, when called upon, to work upon the public roads; but this was peculiarly grievous in Bavaria, where labourers are so scarce that the farmers are frequently obliged to leave a great part of their grounds uncultivated for want of hands.

My plan was to measure all the public roads from the capital cities in the Elector's dominions to the frontiers, and all cross country roads; placing mile-stones regularly numbered upon each road, at regular distances of one hour, or half a German mile from each other;—to divide each road into as many stations as it contained mile-stones; each station extending from one mile-stone to another; and to erect in the middle of each station, by the road-side, a small house, with stabling for three or four horses, and with a small garden adjoining to it;—to place in each of these houses, a small detachment of cavalry of three or four men, —a soldier on furlough, employed to take care of the road and keep it in repair within the limits of the station;—an invalid soldier to take care of the house, and to receive orders and messages in the absence of the others,—to take care of the garden, to provide provisions, and cook for the family.

If any of the soldiers should happen to be married, his wife might have been allowed to lodge in the house, upon condition of her assisting the invalid soldier in this service; or a pensioned soldier's widow might have been employed for the same purpose.

To preserve order and discipline in these establishments, it was proposed to employ active and intelligent non-commissioned officers as overseers of the highways, and to place these under the orders of superior officers appointed to preside over more extensive districts.

It was proposed likewise to plant rows of useful trees by the road-side from one station to another throughout the whole country, and it was calculated that after a certain number of years the produce of those trees would have been nearly sufficient to defray all the expences of repairing the roads.

Such an arrangement, with the striking appearance of order and regularity that would accompany it, could not have failed to interest every person of feeling who saw it; and I am persuaded that such a scheme might be carried into execution with great advantage in most countries where standing armies are kept up in time of peace. The reasons why this plan was not executed in Bavaria at the time it was proposed are too long, and too foreign to my present purpose to be here related. Perhaps a time may come when they will cease to exist.

APPENDIX. No I.

ADDRESS and PETITION to all the Inhabitants and Citizens of MUNICH, in the Name of the real Poor and Distressed.

(Translated from the German).

Too long have the public honour and safety, morality and religion, called aloud for the extirpation of an evil, which, though habit has rendered it familiar to us, always appears in all its horrid and disgusting shapes; and whose dangerous effects show themselves every where, and are increasing every day.

Too long already have the virtuous citizens of this metropolis seen with concern the growing numbers of the Beggars, their impudence, and their open and shameless debaucheries; yet idleness and mendicity (those pests of society) have been so feebly counteracted, that, instead of being checked and suppressed, they have triumphed over those weak attempts to restrain them and acquiring fresh vigour and activity from success, have spread their baleful influence far and wide.

What well-affected citizen can be indifferent to the shame that devolves upon himself and upon his country, when whole swarms of dissolute rabble, covered with filthy rags, parade the streets, and by tales of real or of fictitious distress—by clamorous importunity, insolence, and rudeness, extort involuntary contributions from every traveller? When no retreat is to be found, no retirement where poverty, misery, and impudent hypocrisy, in all their disgusting and hideous forms, do not continually intrude; when no one is permitted to enjoy a peaceful moment, free from their importunity, either in the churches or in public places, at the tombs of the dead, or at the places of amusement? What avail the marks of affluence and prosperity which appear in the dress and equipage of individuals, in the elegance of their dwellings, and in the magnificence and splendid ornaments of our churches, while the voice of woe is heard in every corner, proceeding from the lips of hoary age worn out with labour; from strong and healthy men capable of labour; from young infants and their shameless and abandoned parents? What reputable citizen would not blush, if among the inmates of his house should be found a miserable wretch, who by tales of real or fictitious distress should attempt to extort charitable donations from his friends and visitors? What opinion would he expect would be formed of his understanding—of his heart—of his circumstances? What then must the foreigner and traveller think, who, after having seen no vestige of Beggary in the neighbouring countries, should, upon his arrival at Munich, find himself suddenly surrounded by a swarm of groaning winching wretches, besieging and following his carriage?

THE PUBLIC HONOUR calls aloud to have a stop put to this disgraceful evil.

THE PUBLIC SAFETY also demands it. The dreadful consequences are obvious, which must ensue when great numbers of healthy individuals, and whole families, live in idleness, without any settled abode, concluding every day with schemes for defrauding the public of their subsistence for the next: where the children belonging to this numerous society are made use of to impose on the credulity of the benevolent, and where they are regularly trained, from their earliest infancy, in all those infamous practices, which are carried on systematically, and to such an alarming extent among us.

Great numbers of these children grow up to die under the hands of the executioner. The only instruction they receive from their parents is how to cheat and deceive; and daily practice in lying and stealing from their very infancy, renders them uncommonly expert in their infamous trade. The records of the courts of justice show in innumerable instances, that early habits of Idleness and Beggary are a preparation for the gallows; and among the numerous thefts that are daily committed in this capital, there are very few that are not committed by persons who get into the houses under the pretext of asking for charity.

What person is ignorant of these facts? and who can demand further proofs of the necessity of a solid and durable institution, for the relief and support of the Poor?

The reader would be seized with horror, were we to unveil all the secret abominations of these abandoned wretches. They laugh alike at the laws of God and of man. No crime is too horrible and shocking for them, nothing in heaven or on the earth too holy not to be profaned by them without scruple, and employed with consummate hyprocrisy to their wicked purposes[3].

Whence is it that this evil proceeds? not from the inability of this great capital to provide for its Poor; for no city in the world, of equal extent and population, has so many hospitals for the sick and infirm, and other institutions of public charity. Neither is it owing to the hard-heartedness of the inhabitants; for a more feeling and charitable people cannot be found. Even the uncommonly great and increasing numbers of the Beggars show the kindness and liberality of the inhabitants; for these vagabonds naturally collect together in the greatest numbers, where their trade can be carried on to the greatest advantage.

THE INJUDICIOUS DISPENSATION OF ALMS is the real and only source of this evil.

In every community there are certainly to be found a greater or less number of poor and distressed persons, who have just claims on the public charity. This is also the case at Munich; and nature dictates to us the duty of administering relief to suffering humanity, and more especially to our poor and distressed fellow-citizens; and our Holy Religion promises eternal rewards to him who supports and relieves the poor and needy, and threatens everlasting damnation to him who sends them away without relief.

The Holy Fathers teach, that when there are no other means left for the relief and support of the Poor, the superfluous ornaments of the churches may be disposed of, and even the sacred vessels melted down and sold for that purpose.

But what shall we think, when we see those very persons, who profess to live after the rules and precepts laid down in the word of God, act diametrically contrary to them?

Such, doubtless, is the fatal conduct of those who are induced by mistaken compassion to lavish their alms upon Beggars, and obstruct the relief of the really indigent. — Alms that frustrate a good and useful institution cannot be meritorious, or acceptable to God: and no maxim is less founded in truth, than that the merit of the giver is undiminished by the unworthiness of the object. — The truly distressed are too bashful to mix with the herd of common Beggars; necessity, it is true, will sometimes conquer their timidity, and compel them publicity to solicit charity; but their modest appeal is unheard or unnoticed, whilst a dissolute vagabond, who exhibits an hypocritical picture of distress, — a drunken wretch, who pretends to have a numerous family and to be persecuted by misfortune, — or an impudent unfeeling women, who excites pity by the tears and cries of a poor child whom she has hired perhaps for the purpose, and tortured into suffering, steps daringly forward to intercept the alms of the charitable; and the well-intentioned gift which should relieve the indigent is the prize of impudence and imposition, and the support of vice and idleness. — What then is left for the modest object of real distress, but to retire dispirited and hide himself in the obscurity of his cottage, there to languish in misery, whilst the bolder Beggar consumes the ill-bestowed gift in mirth and riot? And, yet, the charitable donor flatters himself that he has performed an exemplary duty!

We earnestly entreat every citizen and inhabitant of this capital, each in his respective station, no longer to countenance mendicity by such a misapplication of their well-meant charity; contributing thus to augment the fatal consequences of the evil itself, as well as to impede the relief of the real necessitous.

We are firmly persuaded, that by pointing out to our fellow-citizens a method by which they may exercise their benevolence towards the indigent and distressed in a meritorious manner, we shall gratify their pious zeal and humanity, and at the same time essentially promote the honour and safety of the state, and the interests of sound morality and religion.

And this is the sole object of the Military Workhouse, which has been instituted by the command of his Electoral Highness, where, from this time forward, all who are able to work may find employment and wages, and will be cloathed and fed. — THERE will be the really indigent find a secure asylum, and those unfortunate persons who are a prey to sickness and infirmity, or are worn out with age, will be effectually relieved. —

We beg you not to listen to the false representations which may, perhaps, be made to calumniate this institution, by putting it on a level with former imperfect establishments. — Why should not an institution prosper at Munich, which has already been successful in other places, particularly at Manheim, where above 800 persons are daily employed in the Military Workhouse, and heap benedictions on its benevolent founder? — Have the inhabitants of this town less good sense, less humanity, or less zeal for the good of mankind? No — it would be an insult on the patriotism of our fellow-citizens, were we to doubt of their readiness to concur in our undertaking.

The only efficacious way of promoting an institution so intimately connected with the safety, honor, and welfare of the state, and with the interests of religion and morality, is a general resolution of the inhabitants to establish a voluntary monthly contribution, and strictly prohibit the abominable and degrading practice of street-begging; the unlimited exercise of which, notwithstanding its fatal and disgraceful consequences, is perhaps more glaringly indulged in Munich than in any other city in Germany.

In vain will the institution be opposed by the prejudices, or the meanness and malice of persons who are themselves used to mendicity, or to exercise an insolent dominion over Beggars.

It will subsist in spite of all their efforts; and we have the fullest confidence that the generous and well-disposed inhabitants of this city will be sensible how injurious the habits of encouraging public mendicity are, when an opportunity is offered them of contributing to an institution where the really indigent are sure to find assistance, and where the benevolent Christian is certain that his neighbours and fellow-citizens are benefited by his charitable donations.

The simplest and most effectual way of ascertaining the extent of such contribution is to form a list of all the citizens and inhabitants of the town, with the name of the street, and number of the house they inhabit. This register may be called an Alms Book. It will be presented to each inhabitant, that he may put down the sum which he means voluntarily to subscribe every month towards the support of the Poor. The smallest donation will be gratefully received, and the objects who are relieved by them will pray for them to the Almighty Rewarder of all good actions.

As this charitable contribution is to be absolutely voluntary, every one, whatever be his rank or property, will subscribe as he pleases, a greater or a less sum, or none at all. The names of the benefactors and their donations will be printed and published quarterly, that every one may know and acknowledge the zealous friends of humanity, by whose assistance an evil of such magnitude, so long and so universally complained of, will be finally rooted out.

We request that the public will not oppose so sure and effectual a mode of granting relief to the Poor, but rather give their generous support to an undertaking, which cannot but be productive of much good, and acceptable in the sight of Heaven.

To convince every one of the faithful application of these contributions, an exact detail both of the receipt and expenditure of the institution will be printed and laid before the public every three months; and every subscriber will be allowed to inspect and examine the original accounts whenever he shall think proper.

It must be obvious to every one, even to persons of the most suspicious dispositions, that this institution is perfectly disinterested, and owes its origin entirely to pure benevolence, and an active zeal for the public good, when it is known that a Committee appointed by his Electoral Highness, under the direction of the Presidents of the Council of War, the Supreme Regency, and the Ecclesiastical Council, will have the sole administration and direction of the affairs of the institution, and that the monthly collections of alms will be made by creditable persons properly authorised; and that no salary, or emoluments of any kind, will be levied on the funds of the institution, either for salaries for the collectors, or any other persons employed in the service of the institution, as will clearly appear by the printed quarterly accounts. By such precautions, we trust, we shall obviate all possible suspicions, and inspire every unprejudiced person with a firm confidence in this useful institution.

Henceforward, then, the infamous practice of begging in the streets will no longer tolerated in Munich, and the public are from this moment exonerated from a burden which is not less trouble-some to individuals than it is disgraceful to the country. Who can doubt the co-operation of every individual for the accomplishment of so laudable an undertaking? We trust that no one will encourage idleness, by an injudicious and pernicious profusion of alms given to Beggars; and by promoting the most unbridled licentiousness, make himself a participator in the dangerous consequences of men-dicity, and share the guilt of all those crimes and offences which endanger the welfare of the state, injure the cause of religion, and insult the distress of the really indigent.

No longer will these vagabonds impose on good-nature and be-nevolence, by false pretences, by ill-founded complaints of the inef-ficacy of the provision for the Poor, or by any other artifices; nor can they escape the strict and constant vigilance with which they will in future be watched; when every person they meet will direct them to the House of Industry, instead of giving them money.

It is this regulation alone which can effectuate our purpose, a regulation enforced in the days of primitive Christianity, and sanc-tioned by Religion itself; the charitable gifts of the wealthier Chris-tians being in those days all deposited in a common treasury, for the benefit of their poorer and distressed Brethren, and not squandered away in the encouragement of dissolute idleness.

We therefore entreat and beseech the public in general, in the name of suffering humanity, and of that Almighty Being who can-not but regard so laudable an enterprise with an eye of favour, to give every possible support to our design. And we trust that the clergy of every denomination, but especially the public preachers, will exert their splendid abilities to animate their congregations to co-operate with us in this great and important undertaking.

APPENDIX No II.

SUBSCRIPTION LISTS distributed among the Inhabitants of MUNICH, in the Month of JANUARY 1790, when the Establish-ment for the Relief of the Poor in that City was formed.

Translated from the Original German.

VOLUNTARY SUBSCRIPTIONS
for
The Relief and Support of
The Industrious, Sick, and Helpless POOR,
and
For the total Extirpation of VAGRANTS
and STREET-BEGGARS,
In the City of MUNICH.

REMARKS.

These voluntary subscriptions will be collected monthly, namely, on the last Sunday morning of every month, under the direction of the Committee of Governors of the Institution for the Poor; consisting of the President of the Council of War,—the President of the Council of the Regency,—and the President of the Ecclesiastical Council[4]; and the amount of these collections will always be regularly noted down in books kept for that purpose; and at the end of every three months a particular detailed account of the application of these sums will be printed, and given gratis to the subscribers and to the public.

No part of these voluntary contributions will ever be taken, or appropriated to the payment of salaries, gratuities, or rewards to any of those persons who may be employed in carrying on the business of the institution; but the whole amount of the sums collected will be faithfully applied to the relief and support of the Poor, and to that charitable purpose alone, as the accounts of the expenditures of the institution, which will be published from time to time, will clearly show and demonstrate.—All the persons necessary to be employed in the affairs of this establishment, will either be selected from among such as already are in the receipt of salaries, sufficient for their comfortable maintenance from other funds; or they will be such persons, in easy circumstances, as may offer themselves voluntarily for these services, from motives of humanity, and a disinterested wish to be instrumental in doing good.

As the preparations which have been made, and are making for the support of the Poor, leave no doubt, but that adequate relief will be afforded to them in future, they will no longer have any pretext for begging; and all persons are most earnestly requested to abstain henceforward from giving alms to Beggars. Instead of giving money to such persons as they may find begging in the street, they are requested to direct them to the House of Industry, where they will, without fail, receive such assistance and support as they may stand in need of and deserve.

Those persons whose names are already inserted in other lists, as subscribers to this institution, are, nevertheless, requested to enter their names upon these family-sheets; for though their names may stand on several lists, their contributions will be called for upon one of them only, and that one will be the family-sheet.

Those persons of either sex, who have no families, but occupy houses or lodging of their own, are, notwithstanding their being without families, requested to put down the amount of the monthly contributions they are willing to give to this institution upon a family-sheet, and to insert their names in the list as "head of the family."

Under the column destined for the names of "relations and friends, living in the house," may be included strangers, lodgers, boarders, etc.

The column for "domestics" may, in like manner, serve, particularly in the houses of the nobility, and other distinguished persons, for stewards, tutors, governesses, etc.

Each head of a family will receive two of these family-sheets, namely, one with these Remarks, which he will keep for his information,—the other, printed on a half-sheet of paper, and without remarks, which he will please to return to the public office of the institution.

In case of a change in the family, or if one or other of the members of it should think proper to increase or to lessen their contribution, this alteration is to be marked upon the half-sheet, which is kept by the head of the family; and this sheet so altered is to be sent to the public office of the institution, to the end that these alterations may be made in the general lists of the subscribers; or new printed forms being procured from the public office, and filled up, these new lists may be exchanged against the old ones.

For the accommodation of those who may at any time wish to contribute privately to the support of the institution any sums in addition to their ordinary monthly donations, the banker of the institution, Mr. Dallarmi, will receive such sums destined for that purpose, as may be sent to him privately under any feigned name, motto, or device; and for the security of the donors, accounts of all the sums so received, with an account of the feigned name, motto, or device, under which each of them was sent to the banker, will be regularly published in the Munich Gazette.

The first collection will be made on the last Sunday of the present month, and the following collections on the last Monday of every succeeding month; and each head of a family is respectfully requested to cause the contributions of his family, and of the inhabitants of his house, to be collected at the end of every month, by a domestic or a servant, and to keep the same in readiness against the time of the collection.

All persons of both sexes, and of every age and condition, (Paupers only excepted,) are earnestly requested to have their names inserted in these lists or family-sheets; and they may rest assured, that any sum, even the most trifling, will be received with thankfulness, and applied with care to the great object of the institution—the relief and encouragement of the Poor and the Distressed.

And finally, as it cannot fail to contribute very much to improve the human heart, if young persons at an early period of life are accustomed to acts of benevolence,—it is recommended to parents, to cause all their children to put down their names as subscribers to this undertaking, and this, even though the donations they may be able to spare may be the most trifling, or even if the parents should be obliged to lessen their own contributions in order to enable their children to become subscribers.

The foregoing Remarks were printed on the two first pages of a sheet, 13 inches by 18 inches, of strong writing-paper. The following Subscription List was printed on the third page of the same sheet,— and also on a separate half-sheet of the same kind of paper.

Voluntary Contributions for the Support of the Poor at Munich.

F A M I L Y — S H E E T.

========================

Number of the House District Street Floor.
Head of the Family } Monthly Contributions.
His Character, or } Florins. Creutzers.

Other Persons belonging to the Family. — — — — — — — — — — —
— : Wife, Children,
Re- :Monthly :Domestics, Journey- :Monthly : : lations and Friends
:Contribu-:man, Menial Servants, :Contribu-: : of both Sexes living:
tions. :etc of both Sexes, the: tions. : : with the Family. The:
:Christian and Sirname : : : Christian Name and : :of each Individu-
al. : : : Sirname of each Per-:— —:— —: :— —:— —: : son. : Fl.: Kr.: :
Fl.: Kr.: :— — — — — — — — — — —:— —:— —:— — — — — — — — — —
:— —:— —::
::
: : : : : (At the lower corner : : : : : : of this half-sheet : : : : : : was
printed in small : : : : : : : type): "This half- : : : : : : : "sheet is to be sent
: : : : : : : "into the Public : : : : : : : "Office of the : : : : : : : "Institution." : :
: —
— — — —

APPENDIX III.

[Etext editor's note…the following table has had to be split into
two parts, with the additional references A) B) etc through to UK) to
link them together. Originally the entire table was printed in landscape format, with totals carried forward, brought over, which have
been removed.]

An Account of the RECEIPTS and EXPENDITURES of the INSTI-
TUTION for the POOR at MUNICH during Five Years.

R E C E I P T S.

— —
— — — — —

: : : Total in :
: 1790. : 1791. : 1792. : 1793. : 1794. : 5 Years. :
:— — — — —:— — — — —:— — — — —:— — — — —:— — — — —:— —

— — —:
: Florins. : Florins. : Florins. : Florins. : Florins. : Florins. :
: : : : : : : :
A) : 36,640 : 38,024 : 35,847 : 34,424 : 33,880 : 178.815 :
: : : : : : :
B) : 15,400 : 15,400 : 16,800 : 16,800 : 16,800 : 81,200 :
: : : : : : :
C) : 970 : 1,043 : 800 : 800 : 802 : 4,415 :
: : : : : : :
D) : 179 : 388 : 388 : 411 : 390 : 1,756 :
: : : : : : :
E) : — — — : 168 : 392 : 229 : 234 : 1,023 :
: : : : : : :
F) : — — — : — — — : — — — : 3,216 : 2,773 : 5,989 :
: : : : : : :
G) : 318 : 177 : 187 : 610 : 229 : 1,521 :
: : : : : : :
H) : 99 : 153 : 69 : 168 : 176 : 665 :
: : : : : : :
I) : 3,642 : 691 : 825 : 723 : 423 : 6,304 :
: : : : : : :
J) : 2,674 : 1,472 : 3,528 : 1,820 : 12,179 : 21,673 :
: : : : : : :
K) : 48 : 128 : 48 : 48 : — — — : 272 :
: : : : : : :
L) : 3,300 : 4,600 : 1,500 : — — — : — — — : 9,400 :
: : : : : : :
M) : 824 : 3,433 : 910 : 1,752 : 346 : 7,265 :
:==========:==========:==========:==========:======
====:==========:
: 64,094 : 65,677 : 61,294 : 61,001 : 70,232 : 320,298 :

EXPENDITURES.

— —
— — — —-
: : : Total in :
: 1790. : 1791. : 1792. : 1793. : 1794. : 5 Years. :

:— — — — —:— — — — —:— — — — —:— — — — —:— — — — —:— —
— — —:
: Florins. : Florins. : Florins. : Florins. : Florins. : Florins. :
: : : : : : :
N) : 42,080 : 46,410 : 43,055 : 41,933 : 43,189 : 216,667 :
: : : : : : :
O) : 11,800 : 9,900 : 10,300 : 9,600 : 9,400 : 51,000 :
: : : : : : :
P) : 1,011 : 1,040 : 800 : 861 : 805 : 4,517 :
: : : : : : :
Q) : 450 : 403 : 350 : 1,150 : 1,500 : 3,853 :
: : : : : : :
R) : 217 : 254 : 272 : 336 : 290 : 1,396 :
: : : : : : :
S) : 256 : 183 : 219 : 210 : 226 : 1,094 :
: : : : : : :
TA): 890 : 564 : 418 : 425 : 594 : 2,891 :
: : : : : : :
TB): 160 : 187 : 34 : 35 : 94 : 510 :
: : : : : : :
TC): 960 : 960 : 960 : 960 : 960 : 4,800 :
: : : : : : :
TD): 84 : 72 : 72 : 72 : 72 : 372 :
: : : : : : :
TE): 100 : 360 : 288 : 540 : 300 : 1,588 :
: : : : : : :
TF): 220 : 240 : 240 : 240 : 240 : 1,180 :
: : : : : : :
TG): 480 : 480 : 480 : 480 : 480 : 2,400 :
: : : : : : :
TH): 440 : 480 : 480 : 480 : 480 : 2,360 :
: : : : : : :
UA): 318 : 318 : 159 : — — — : — — — : 795 :
: : : : : : :
UB): — — — : — — — : — — — : 183 : 200 : 383 :
: : : : : : :
UC): 1,672 : 1,824 : 912 : — — — : — — — : 4,408 :
: : : : : : :
UD): 369 : 199 : 189 : 250 : 361 : 1,368 :

: : : : : : : :
UE): 506 : 333 : 150 : 227 : 301 : 1,517 :
: : : : : : : :
UF): 22 : 6 : — — — : — — — : — — — : 28 :
: : : : : : : :
UG): 55 : 60 : 60 : 50 : 75 : 300 :
: : : : : : : :
UH): 831 : 300 : — — — : — — — : — — — : 1,131 :
: : : : : : : :
UI): — — — : — — — : 40 : 40 : 40 : 120 :
: : : : : : : :
UJ): — — — : — — — : — — — : — — — : 1,200 : 1,200 :
: : : : : : : :
UK): 172 : 234 : 261 : 645 : 433 : 1,745 :
:==========:==========:==========:==========:======
====:==========:
: 63,093 : 64,807 : 59,739 : 58,717 : 61,240 : 307,596 :
— —
— — — —-

RECEIPTS.

A) From monthly voluntary donations of the inhabitants including 100 Florins given monthly by his Most Serene Highness the Elector out of his private purse; 50 florins monthly by the Electress Dowager of Bavaria, and 50 florins monthly by the States of Bavaria,

B) From the Public Treasury a stated monthly allowance, intended principally to defray the expense of the police of the city,

C) From voluntary donations, particularly destined by the donors to assist the Poor in paying their house-rent,

D) From voluntary and unsolicited donations from the foreign merchants and traders assembled at Munich at the two annual fairs,

E) From the courts of justice, being fines for certain petty offences,

F) From the magistrates of the city; being the amount of sums received from musicians for licence to play in the public houses,

G) From the poor's boxes in the different churches,

H) From the poor's boxes at inns and taverns,

I) From private contributions sent to the banker of the Institution, under feigned names, devices, etc.

J) From legacies,

K) From interest of money due to the Institution,

L) From cash received in advance,

M) From sundries,

EXPENDITURES.

N) Given to the Poor in alms, in ready money,

O) Expended in feeding the Poor at the Public Kitchen of the Military
Workhouse, and in premiums for the encouragement of industry,

P) Given to the Poor to assist them in paying their house-rent,

Q) Paid for medicines administered to the Poor at their own lodgings,

R) Expended in burials,

S) Given with poor children when bound apprentices,

Given as an indemnification for the loss of the right formerly enjoyed of making collections of alms among the inhabitants:

— — —- TA) To persons who have suffered by fires,
— — —- TB) To travelling journeymen tradesmen,
— — —- TC) To the sisters of the religious order of charity,
— — —- TD) To the nuns of the English convent,
— — —- TE) To the hospital for lepers on the Gasteig,
— — —- TF) To the hospital at Schwabing,
— — —- TG) To the poor scholars of the German school,
— — —- TH) To the poor scholars of the Latin school,

UA) Paid to the clerks of office of police

UB) Paid to the accountant of the Institution,

UC) Paid to the guards of the police[5],

UD) Paid to writers employed occasionally as clerks,

UE) Paid to printers and bookbinders,

UF) Paid to the soldiers of the garrison for arresting Beggars,

UG) Gratuities to the schoolmaster at Charles's Gate,

UH) Paid various sums due from the Institution,

UI) Paid interest of monies due,

UJ) Money advanced for purchasing grain,

UK) Sundries,

APPENDIX, No IV.

Certificate relative to the EXPENCE of FUEL in the Public Kitchen of the Military Workhouse at MUNICH

We whose Names are underwritten certify, that we have been present frequently when experiments have been made to determine the expence of Fuel in cooking for the Poor in the Public Kitchen of the Military Workhouse at Munich; and that when the ordinary dinner has been prepared for ONE THOUSAND persons, the expense for Fuel has not amounted to quite twelve creutzers (less than 4 1/2d. sterling).

Baron de Thibout, Heerdan,
   Colonel. Councillor of War.

Munich, 1st September 1795.

APPENDIX, No V.

Printed Form for the DESCRIPTIONS of the POOR.

Description of the poor Person, No

Name

Described Munich, the th of 179

=======================================

Age Years. Stature Feet Inches

Bodily Structure Hair

Eye Complexion

Bodily Defects

Other particular Marks

State of Health

Place of Nativity

Lives here since

Came here from In what Manner

Profession Religion

Quality Family

Supports himself, at present, by

Lives at present Quarter, District, Street,

House, No Floor,

Can be considered as a Pauper belonging to this City, and ought therefore to be

Is capable of doing the following Work:

Could be trained to the following Occupations:

:fl.:kr.:

Could gain by this Work per Week — — : : :
Wants for his weekly Support — — — — : : :
Receives at present per Week from his own } : : :
 Means, get by way of Pension, Alms, } : : :
and .. .. } : : :
Wants, therefore, a weekly Allowance of Alms of : : :

: : :

— — — — —-

:fl.:kr.:

{ Income of his own — — : : :
{ Earned by working — — : : :
{ Salary — — — — : : :
Enjoyed heretofore { Pension — — — — : : :
per Week { { From the Court : : :
{ Alms, { From the City — : : :

{ { From private Persons : : :
{ Got by begging — — : : :
: : :
:—-:—-:
Total : : :
: : :
— — — —-

:fl.:kr.:
: : :
Pays House-rent — — — — — — — — — : : :
: : :
Has Bed of his own, the Value of which : : :
is about— — — — — — — — — — — : : :
: : :
Possesses other Utensils necessary for House- : : :
keeping, worth about— — — — — — — — : : :
: : :
Is provided with the following Working Tools: — : : :
: : :
— — — —-

Can work at Home

Could be employed in the Military Workhouse

Is provided with Raiment, and wants

Articles of Apparel

Life and Conduct, according to the Information received

Is given to and

Is known to have committed Crimes and has appeared before the Magistrates

How long he lives in his present Habitation
    Year Month Weeks

Name and Residence of his present Landlord

Where he lived before, and how long

Other Remarks.

Has been settled here

Received a Licence to marry, from

Possessed or received, when married
                    Value about fl. kr.

Was reduced to Poverty by

Is poor and in want, since

Could not extricate himself from his Difficulties, because

N.B. This Form is printed on a Half-sheet of strong
Writing Paper, folded together so as to make two Leaves in
Quarto; each Leaf being 8 Inches high, and 6 1/2 Inches wide.

APPENDIX, No VI.

Printed Form for SPIN-TICKETS, such as are used at the Military
Workhouse at Munich.

Munich Military Workhouse,
179 the No
        received
 lb. of
Delivered back skains knots
of weighing lb. oz.
Is entitled to receive per xrs.
Total,

Attest. this 179

This printed Form is filled up as follows:

Munich Military Workhouse, 1795 the 1st Sept. No 134. Mary Smith received 1 lb. of Flax, No 3, Delivered back 2 skains 3 knots of Thread, weighing 1 lb. —- oz. Is entitled to receive per lb. xrs. 10. Total, ten creutzers. Attest. this 4th Sept. 1795

Will Wildmann.

An improved Form for a Spin-Ticket, with its Abstract; which Abstract is to be cut off from the Ticket, and fastened to the Bundle of Yarn or Thread.

– –

– – – : Spin-Ticket. :: Abstract of : : Munich House of Industry. :: Spin-Ticket. : : 1795 the 10th Sept. No 230. :: Munich House : : Mary Smith received :: of : : 1 lb. of wool, No 14. :: Industry, : : Delivered back 2 skains 4 knots :: 1795, the 10th Sept. : : of yarn, weighing 1 lb. – oz. :: No 230. : : Wages per lb. for spinning 12 xrs. :: 2 skains 4 knots : : Is entitled to receive twelve xrs. :: of woollen yarn, : : Attest. this 14th of Sept. 1795. :: Spinner, Mary Smith. : : J. Schmidt. :: Attest. J. Schmidt. : : :: : –

– – – – – – – – – –

In order that the original entry of the Spin-Tickets in the general tables, kept by the clerks of the Spinners, may more readily be found, all the Tickets for the same material, (flax, for instance,) issued by the same clerk, during the course of each month, must be regularly numbered.

APPENDIX, No VII.

An Account of EXPERIMENTS made at the BAKE-HOUSE of the MILITARY WORKHOUSE at MUNICH, November the 4th and

5th, 1794.

In baking RYE BREAD

The oven, which is of an oval form, is 12 feet deep, measured from the mouth to the end; 11 feet 10 inches wide, and 1 foot 11 inches high, in the middle.

November 4th, at 10 o'clock in the morning, 1736 lbs.[6] of rye meal were taken out of the store room, and sent to the bakehouse, where it was worked and baked into bread, at six different times, in the following manner: —

FIRST BATCH

At 45 minutes after 10 o'clock, the meal was mixed for the first time, for which purpose 16 quarts (Bavarian measure) of lukewarm water, weighing 28 lbs. 28 loths, were used.

At 3 o'clock in the afternoon, the little leaven (as it is called) was made, for which purpose 24 quarts, or 43 lbs. 10 loths of water were used; and at half an hour after 7 o'clock, the great leaven was made with 40 quarts, or 72 lbs. 6 loths, of water. At 11 o'clock this mass was prepared for kneading, by the addition of 40 quarts, or 72 lbs. 6 loths, more of water.

At 15 minutes after 10 o'clock at night, the kneading of the dough was commenced; 2 1/2 lbs. of salt being first mixed with the mass. The dough having been suffered to rise till a quarter before 2 o'clock, it was kneaded a second time, and then made, in half an hour's time, into 191 loaves, each of them weighing 2 lbs. 16 loths. These loaves having been suffered to rise half an hour, they were put into the oven 10 minutes before 3 o'clock, and in an hour after taken out again, when 25 loaves being immediately weighed, were found to weight 55 lbs. 15 loths. Each loaf, therefore, when baked, weighed 2 lbs. 5 1/2 loths; and as it weighed 2 lbs. 16 loths when it was put into the oven, it lost 10 1/2 loths in being baked.

The whole quantity of water used in this experiment, in making the leaven and the dough, was 216 lbs. 18 loths. — The quantity of meal used was about 310 lbs.

First Heating of the Oven

This was begun 35 minutes after four o'clock, with 220 1/2 lbs. of pine-wood, which was in full flame 15 minutes after five o'clock. — At 8 minutes after 8 o'clock, 51 lbs. more of wood were added; — 12 minutes after 11 o'clock, 32 lbs. more were put into the oven; — 51 lbs. at one o'clock, and 12 lbs. more at 30 minutes after 2 o'clock; so that 366 lbs. 16 loths of wood were used for the first heating.

SECOND BATCH.

At 20 minutes after 11 o'clock, the proper quantity of leaven was mixed with the meal, and 44 quarts, or 79 lbs. 25 loths, of water added to it. At 10 minutes after 3 o'clock, the meal was prepared for kneading, by adding to it 52 quarts, or 93 lbs. 27 loths, of water.

At 30 minutes after 5 o'clock, the kneading of the dough was begun; 2 1/2 lbs. of salt having been previously added. At 15 minutes after 6 o'clock, the dough was kneaded a second time, and formed into 186 loaves, which were put into the oven at 15 minutes after 7 o'clock, and taken out again 9 minutes after 8 o'clock, when 25 loaves being immediately weighed, were found to weigh 55 lbs. 4 loths. — Water used in making the second dough, 173 lbs. 8 loths.

Second Heating of the Oven

This was begun 20 minutes after 4 o'clock in the morning, with 54 1/2 lbs. of wood; 20 lbs. were added 10 minutes after 5 o'clock, and 60 lbs. more 6 minutes after 6 o'clock; so that the second heating of the oven required 134 lbs. 16 loths of wood.

THIRD BATCH

At 20 minutes after 3 o'clock, the proper quantity of leaven was mixed with the meal, and 48 quarts, or 86 lbs. 20 loths, of water were put to it.

At 6 minutes after 8 o'clock, this mass was prepared for kneading, by adding to it 48 quarts, or 86 lbs. 20 loths, of water. — At 30 minutes after 9 o'clock, this dough was mixed with 2 1/2 lbs. of salt; and at 30 minutes after 10 o'clock, it was made into 189 loaves, which, after having been suffered to rise for half an hour were put into the oven 10 minutes after 11 o'clock, and taken out again at 12 o'clock.

Fifty loaves of bread, which were weighed immediately upon their being taken out of the oven, were found to weigh 110 lbs. 30 loths; which gives 2 lbs. 5 1/2 loths for the weight of each loaf. The water used in making this batch of bread was 173 lbs. 8 loths.

Third Heating of the Oven.

This was begun 30 minutes after 8 o'clock, with 50 lbs. of wood; and 50 lbs. more being added 30 minutes after 9 o'clock, the whole quantity used was 100 lbs.

FOURTH BATCH.

At a quarter before 8 o'clock, the proper quantity of leaven was mixed with the meal, and 48 quarts, or 86 lbs. 20 loths, of water being added, at 30 minutes past 11 o'clock, this mass was prepared for kneading, by adding to it 52 quarts, or 93 lbs. 27 loths, of water.

Four minutes after 1 o'clock, 2 1/2 lbs. of salt were added. The dough being kneaded at 15 minutes after two o'clock, 188 loaves of bread were made, which were put into the oven 5 minutes before 3 o'clock, and taken out again at the end of one hour, when 25 of them were weighed, and found to weigh, one with the other, 2 lbs. 5 1/2 loths.

The water used in making this batch of bread was 180 lbs. 15 loths.

Fourth Heating of the Oven.

This was begun 15 minutes after 12 o'clock, with 40 lbs. of wood, and 50 lbs. more being added at 30 minutes after 1 o'clock, the total quantity used was 90 lbs.

FIFTH BATCH.

At 1/4 before 12 o'clock, the proper quantity of leaven was mixed with the meal, and 52 quarts, or 93 lbs. 27 loths, of water put into it. — This mass was prepared for kneading at 15 minutes after 4 o'clock, by the addition of 48 quarts, or 86 lbs. 20 loths, of water. The kneading of the dough was begun at 5 o'clock, and at 30 minutes after 5 it was made into loaves, 2 1/2 lbs. of salt having been previ-ously added. 186 loaves being made out of this dough, they were put into the oven at 10 minutes before 7 o'clock, and taken out again at the end of one hour, when 25 loaves were weighed, and found to weigh 55 lbs. 18 loths.—The quantity of water used in making the dough for this batch of bread was 180 lbs. 15 loths.

Fifth Heating of the Oven

The oven was begun to be heated the fifth time at 15 minutes after four o'clock, with 40 lbs. of wood, and 40 lbs. more were added at 6 o'clock; so that in this heating no more than 80 lbs. of wood were consumed.

SIXTH BATCH.

The meal was mixed with leaven at 30 minutes after 3 o'clock; for which purpose 32 quarts, or 57 lbs. 24 loths, of water were used at 15 minutes after 7 o'clock. This mass was prepared for kneading, by the addition of 44 quarts, or 79 lbs. 13 loths, of water, and a proportion of salt; at 19 minutes after 9 o'clock the dough was kneaded the first, and at 1/4 before 10 the second time; and in the course of half an hour 160 loaves were made out of it, which were put into the oven at 10 minutes before 11 o'clock, and taken out again at 8 minutes before 12 o'clock at midnight.

The water used in making the dough for this batch of bread was 137 lbs. 5 loths.

Sixth Heating of the Oven.

At 1/4 after 8 o'clock, the sixth and last fire was made with 40 lbs. of wood; to which, at 15 minutes before 10 o'clock at night, 34 1/2 lbs. more were added; so that in the last heating 74 1/2 lbs. of wood only were consumed.

GENERAL RESULTS of these EXPERIMENTS.

The ingredients employed in making the bread in these six experiments were as follows: viz.

```
                lbs. loths.
Of rye meal, — — 1736 0
Of water,— — — 1061 5
```

Of salt, − − − 15 0
______

In all, 2812 5 in weight.

Of this mass 1102 loaves of bread were formed, each of which, before it was baked, weighed 2 1/2 lbs.; consequently, these 1102 loaves, before they were put into the oven, weighed 2755 lbs.: but the ingredients used in making them weighed 2812 lbs. 5 loths. Hence it appears, that the loss of weight in these six experiments, in preparing the leaven, — from evaporation, before the bread was put into the oven, — from waste, etc. — amounted to no less than 57 lbs. 5 loths.

In subsequent experiments, where less water was used, this loss appeared to be less by more than one half.

In these experiments 1061 lbs. 5 loths of water were used to 1736 lbs. of meal, which gives 61 lbs. 4 3/4 loths of water to 100 lbs. of meal. But subsequent experiments showed 56 lbs. of water to be quite sufficient for 100 lbs. of the meal.

These 1102 loaves, when baked, weighed at a medium 2 lbs. 5 1/2 loths each; consequently, taken together, they weighed 2393 lbs. 13 loths: and as they weighed 2755 lbs. when they were put into the oven, they must have lost 361 lbs. 19 loths in being baked, which gives 10 1/2 loths, equal to 21/160 or nearly 1/8 of its original weight before it was baked, for the diminution of the weight of each loaf.

According to the standing regulations of the baking business carried on in the bakehouse of the Military Workhouse at Munich, for each 100 lbs. of rye meal which the baker receives from the storekeeper, he is obliged to deliver 139 lbs. of well-baked bread; namely, 64 loaves, each weighing 2 lbs. 5 1/2 loths. And as in the before-mentioned six experiments, 1736 lbs. of meal were used, it is evident that 1111 loaves, instead of 1102 loaves, ought to have been produced; for 100 lbs. of meal are to 64 loaves as 1736 lbs, to 1111 loaves. Hence it appears that 9 loaves less were produced in these experiments than ought to have been produced.

There were reasons to suspect that this was so contrived by the baker, with a design to get the number of loaves he was obliged to deliver for each 100 lbs. of meal lessened;—but in this attempt he did not succeed.

Quantity of Fuel consumed in these Experiments.

Dry pine-wood.
lbs. loths.

In heating the oven first time, — — 366 16
        second time, — — 134 16
        third time, — — 100 0
        fourth time, — — 90 0
        fifth time, — — 80 0
        sixth time, — — 74 16

— — — — — —

Total, 845 16

Employed in keeping up a small fire
near the mouth of the oven while the
bread was putting into it, — — — 34 16

Total consumption of wood in the six experiments, — — — — — — — — 880 lbs.

The results of these experiments show, in a striking manner, how important it is to the saving of fuel in baking bread, to keep the oven continually going, without ever letting it cool: for in the first experiment when the oven was cold, when it was begun to be heated, the quantity of wood required to heat it was 366 1/2 lbs.; but in the sixth experiment, after the oven had been well warmed in the preceding experiments, the quantity of fuel required was only 74 1/2 lbs.

As in these experiments 2393 lbs. 13 loths of bread were baked with the heat generated in the combustion of 880 lbs. of wood, this gives to each pound of bread 11 1/3 loths, or 34/96 of a pound, of wood.

In the fifth experiment, or batch, 186 loaves weighing (at 2 lbs. 5 1/2 loths each) 304 lbs. were baked, and only 80 lbs. of wood consumed, which gives but a trifle more than 1/4 of a pound of wood to each pound of bread; or 1 pound of wood to 4 pounds of bread.

As each loaf weighed 2 lbs. 16 loths when it was put into the oven, and only 2 lbs. 5 1/2 loths when it came out of it, the loss of weight each loaf sustained in being baked was 10 1/2 loths, as has already been observed. Now this loss of weight could only arise from the evaporation of the superabundant water existing in the dough; and as it is known how much heat, and consequently HOW MUCH FUEL is required to reduce any given quantity of water, at any given temperature, to steam, it is possible, from these data, to determine how much fuel would be required to bake any given quantity of bread, upon the supposition that NO PART OF THE HEAT GENERATED IN THE COMBUSTION OF THE FUEL WAS LOST, either in heating the apparatus, or in any other way; but that the whole of it was employed in baking the bread, and in that process alone. And though these computations will not show how the heat which is lost might be saved, yet, as they ascertain what the amount of this loss really is in any given case, they enable us to determine, with a considerable degree of precision, not only the relative merit of different arrangements for economizing fuel in the process of baking, but they show also, at the sane time, the precise distance of each from that point of perfection, where any farther improvements would be impossible: And on that account, these computations are certainly interesting.

In computing how much heat is NECESSARY to bake any given quantity of bread, it will tend much to simplify the investigation, if we consider the loaf as being first heated to the temperature of boiling water, and then baked in consequence of its redundant water being sent off from it in steam.

But as the dough is composed of two different substances, viz. rye meal and water, and as these substances have been found by experiment to contain different quantities of absolute heat; or, in other words, to require different quantities of heat, to heat equal quantities or weights of them to any given temperature, or any given number of degrees, it will be necessary to determine how much of each of the ingredients is employed in forming any given quantity of dough.

Now, in the foregoing experiments, as 1102 loaves of bread were formed of 1736 lbs. of rye meal, it appears, that there must have been 1.47 lb. of the meal in each loaf; and as these loaves weighed 2 1/2 lbs. each when they were put into the oven, each of them must, in a state of dough, have been composed of 1.47 lb. of rye meal, and 1.03 lb. of water.

Supposing these loaves to have been at the temperature of 55 degrees of Fahrenheit's Thermometer when they were put into the oven, the heat necessary to heat one of them to the temperature of 212 degrees, or the point of boiling water, may be thus computed.

By an experiment, of which I intend hereafter to give an account to the Public, I found, that 20 lbs. of ice-cold water might be made to boil, with the heat generated in the combustion of 1 lb. of dry pine-wood, such as was used in baking the bread in the six experiments before mentioned. Now, if 20 lbs. of water may be heated 180 degrees, (namely from 32 to 212 degrees,) by the heat generated in the combustion of 1 lb. of wood, 1.03 lb. of water may be heated 157 degrees, (from 55 degrees, or temperate, to 212 degrees,) with 0.4436 of a pound of the wood.

Suppose now that rye meal contained the same quantity of absolute heat as water,—as the quantity of meal in each loaf, was 1.47 lb., it appears, that this quantity would have required, (upon the above supposition,) to heat it from the temperature of 55 degrees, to that of 212 degrees; a quantity of heat equal to that which would be generated in the combustion of 0.06405 of a pound of the wood in question.

But it appears, by the result of experiments published by Dr. Crawford, that the quantities of heat required to heat any number of degrees, the same given quantity (in weight) of water and of wheat, (and it is presumed, that the specific or absolute heat of rye cannot be very different from that of wheat,) are to each other, as 2.9 to 1,—water requiring more heat to it, than the grain in that proportion: Consequently, the quantity of wood required to heat from 55 to 212 degrees, the 1.47 lb. of rye meal which entered into the composition of each loaf, instead of being .06405 of a pound, as above determined, upon the false supposition that the specific heat of water and that of rye were the same, would, in fact, amount to no more than 0.02899; for 2.9 (the specific heat of water) is to 1 (the specific heat of rye), as 0.06405 is to 0.02899.

Hence it appears, that the wood required as fuel to heat (from the temperature of 55 degrees to that of 212 degrees) a loaf of rye bread (in the state of dough), weighing 2 1/2 lbs., would be as follows, namely:

Of pine-wood, To heat 1.03 lb. of water, which enters into the composition of the dough, .. 0.04436

To heat the rye meal, 1.47 lb in weight, .. 0.02899

— — — —

Total, 0.07335 lb.

To complete the computation of the quantity of fuel necessary in the process of baking bread, it remains to determine, how much heat is required, to send off in steam, from one of the loaves in question (after it has been heated to the temperature of 212 degrees), the 10 1/2 loths, equal to 21/64 of a pound of water, which each loaf is known to lose in being baked.

Now it appears, from the result of Mr. Watt's ingenious experiments on the quantity of latent heat in steam, that the quantity of heat necessary to change any given quantity of water ALREADY BOILING HOT to steam, is about five times and a half greater than would be sufficient to heat the same quantity of water, from the temperature of freezing, to that of boiling water.

But we have just observed, that 20 lbs. of ice-cold water may be heated to the boiling point, with the heat generated in the combustion of 1 lb. of pine-wood; it appears therefore that 20 lbs. of boiling water would require 5 1/2 times as much, or 5 1/2 lbs. of wood to reduce it to steam.

And if 20 lbs. of boiling water require 5 1/2 lbs. of wood, 21/64 of a pound of water boiling hot will require 0.09023 of a pound of wood to reduce it to steam.

If now, to this quantity of fuel, — — 0.09023 lb. we add that necessary for heating the loaf to the temperature of boiling water, as above determined, — — — 0.07335 lb. — — — — this gives the total quantity of fuel necessary for baking one of these loaves of bread, — — — — — — — — — 0.16353 lb.

Now as these loaves, when baked into bread, weighed 2 lbs. 5 1/2 loths = 2 11/64 lbs. each and required, in being baked, the consumption of 0.16353 of a pound of wood, this gives for the expence of fuel in baking bread 0.07532 of a pound of pine-wood to each pound of rye bread; which is about 13 1/4 lbs. of bread to each pound of wood.

But we have seen, from the results of the before-mentioned experiments, that when the bread was baked under circumstances the most favourable to the economy of fuel, no less than 80 lbs. of pine-wood were employed in heating the oven to bake 304 lbs. of bread, which gives less than 4 lbs. of bread to each pound of wood; consequently, TWO THIRDS at least of the heat generated in the combustion of the fuel must, in that case, have been lost; and in all the other experiments the loss of heat appears to have been still much greater.

A considerable loss of heat in baking will always be inevitable; but it seems probable, that this loss might, with proper attention to the construction of the oven, and to the management of the fire, be reduced at least to one half the quantity generated from the fuel in its combustion. In the manner in which the baking business is now generally carried on, much more than three quarters of the heat generated, or which might be generated from the fuel consumed, is lost.

APPENDIX, No VIII.

The following Account of the Persons in the House of Industry in Dublin, the 30th of April 1796, and of the Details of the Manner and Expence of feeding them, was given to the Author, by order of the Governors of that Institution.

Average of the Description of Poor for the Week ending 30th of April 1796. Males. Females. Total. Employed — — — — 74 352 426 Infirm and Incurable — 172 585 757 Idiots — — — — 16 13 29 Blind— — — — — 5 10 16 — —- — —- — —- 267 960 1227 In the Infirmary. Sick Patients, Servants, etc. 88 200 } }— 343 Lunaticks— — — — 15 40 } — — — Total 1570

Employed at actual labour 322 Persons.
Ditto at menial offices 104 ditto
————-
Total 426

Amongst the 1570 Persons above mentioned, are 282 Children and 447 compelled Persons.

Of the Children, 205 are taught to spell, read, and write.

Saturday, April 30, 1796.

1227 Persons fed at Breakfast.

120 Servants in New-House,
    a 8 oz. bread — — — 60 } lbs. loaves lb. value.
336 Incurables, Children, etc. } 186 is 41 1 1/2 L. 1 14
    a 6 ditto — — — —126 }
771 Workers, etc. got Stirabout.
————-
1227

Weight of meal for Stirabout 4 cwt. costs L. 3 1 8

120 Servants in New-House }
    get 1 quart butter-milk Gal. P.}
    each 30 0 } 167 gallons of
1084 Workers, Incurables, etc. } butter-milk
    1 pint ditto 135 4 } value 1 L.
  23 Sucklers get no butter-milk }
————- Allowed for waste — — 1 4 }
1227

Brought down, L. 5 15 8
s. d.
Fuel to cook the Stirabout, 3 bush. cost 2 3 }
            }0 3 0 1/2
Salt for ditto, 1 qr. 3 lb cost— — 0 9 1/2 }
————————-

The Breakfast cost L. 5 18 8 1/2

Quantity of water, 5 barrels 6 gallons.

1227 Persons fed at Dinner.—BREAD and MEAL POTTAGE. 120 Servants a 9 oz. — 68 } bread } lbs. loaves. lb. value. 1107 Workers, Incurables, } 621 1/2 is 138 0 1/2 L. 5 10 4 etc. 8 oz. ditto—553 1/2} Weight of meal for the pottage, 1 cwt. 3 qrs.— — — 0 13 5 Pepper for ditto, half a pound — — — — — — 0 1 1 Ginger for ditto, 1 pound — — — — — — — 0 1 3 Salt for ditto, 21 pound — — — — — — — — 0 0 7 Fuel for ditto, 3 bushels 2 pecks— — — — — — 0 2 7 1/2 — — — — — — —- Dinner cost L. 6 9 3 1/2

SUPPER.

For 165 Sickly Women on 6 oz, bread. 62 } lbs. loaves lb. value.
   251 Children, 3 oz. do. 47 } 109 is 24 1 0 19 11

   N.B. The expenses of Food for the Hospital, in which there are 343 persons, is not included in the above account.

   Sunday, May 1, 1796. 1220 Persons fed at Breakfast.

120 Servants, a 8 oz. bread.
330 Incurables, Children, etc. 6 oz. do.
770 Workers, etc. get Stirabout.
— —-
1220 Persons.

   The same quantity of provisions delivered this day for Breakfast as on Saturday, and cost the same: viz. 5L. 18s. 8 1/2d.

1220 Persons fed at Dinner. — BREAD, BEEF and BROTH.
                    Cost
 120 Servants, a 9 oz. bread, 68 } lbs. loaves lbs. L. s. d.
1100 Workers, Incurables, etc. } 618 is 137 1 1/2 5 9 6
    8 do. — — — — — 550 }
— —-
1220 Persons.
               Cwt. qrs. lbs.
Weight of raw beef, 4 2 10
Allowed for bone, 1 0 0

           — — — — — — —-
         5 2 10 — 7 19 3
Meal for the broth, 1 2 0 — 1 3 1 1/2
Waste bread for do. 1 0 0 — 0 0 0
Salt for do. 0 0 24 — 0 0 8

Pepper for do. 0 0 0 1/2 — 0 1 1
Fuel, 4 bushels 2 pecks, — 0 3 4 1/2

— — — — — — — —-

Total L. 14 17 0

SUPPER.

The same number of women and children as yesterday, and the Supper cost the same: viz. 19s. 11d.

Wednesday, May 4, 1796.

1216 Persons fed at Breakfast.

120 Servants in New-House, a 8 oz. bread
334 Incurables, Children, etc. a 6 oz. do.
762 Workers, etc. get Stirabout.
— —-
1216 Persons.

The same quantity of provisions, etc. delivered this day for Break-fast as for Saturday, and cost the same: viz. 8L 18s. 8 1/2d.

1216 Persons fed at Dinner. — CALECANNON and BEER.

Cost.

Weight of raw potatoes Cwt. qrs. lbs. L. s. d.
 for Calecannon, — — 19 0 0 — 3 6 6
An allowance for waste, 1 0 0

— — — — — — —-

Weight used, 18 0 0 —
Raw greens for ditto, — 8 0 0 — 1 6 0
Butter for ditto, — — 1 0 0 — 3 12 0
Pepper for ditto, — — 0 0 0 1/2 — 0 1 1
Ginger for ditto, — — 0 0 1 — 0 1 3
Onions for ditto, — — 0 0 14 — 0 2 0
Salt for ditto, — — 0 0 24 — 0 0 8
Fuel, 4 bushels 2 pecks, — 0 3 4
 Time of boiling about four hours.

1193 Persons get 1 }
   pint of beer Galls. p. } Barrs.
   each, making 149 1 }Galls. Galls.
  23 On the breast } 151 is 3 31 2 5 3
— — get no beer. }
1216 }
   Allowed for }
    waste, — 1 7 }

Bread to Incurables and Children on the
breast, 43 loaves, — — — — — — — — 1 15 4
                    — — — — — —-
              Total L. 12 13 5

SUPPER.

The same number of Women and Children as on
Saturday, and cost the same: viz. 19s 11d.

   N.B. All these accounts are in avoirdupois weight, and Irish money.

   APPENDIX, No IX.

   An Account of an EXPERIMENT made (under the Direction
   of the AUTHOR) in the Kitchen of the HOUSE of INDUSTRY
   at DUBLIN, in COOKING for the POOR.

May the 6th, 1796, a dinner was provided for 927 persons of
Calecannon, a kind of food in great repute in Ireland, composed of
Potatoes, boiled and mashed, mixed with about one-fifth of their
weight of boiled Greens, cut fine with sharp shovels, and seasoned
with butter, onions, salt, pepper, and ginger. The ingredients were
boiled in a very large iron boiler, of a circular, or rather hemispheri-
cal form, capable of containing near 400 gallons, and remarkably
thick and heavy. 273 gallons of pump water were put into this boiler; and the following Table will show, in a satisfactory manner, the
progress and the result of the experiment:

Heat Contents of the Boiler
Fuel laid of the
Time. on Coals. Liquid Quantity
Pecks Weight Ingredients. Gall. lbs.
7h 48m 4 106 lb. 55 Water to boil 273
8h 15m 1 26 1/2 the Greens
40m 1 26 1/2 and Potatoes
9h 0m 1 26 1/2
15m 2 53 80
30m 1 26 1/2 90
45m 2 53 110
10h 0m 1 26 1/2 150
20m 212 The Greens
were now put 295 1/2
in.
2m 180
30m 1 26 1/2 190
45m 212
11h the Greens
taken out and 1615
Potatoes put
in.
11h 10m 2 53 180
20m 1 26 1/2 200
30m 212
45m Potatoes done.

GENERAL RESULTS of the EXPERIMENT.

The fuel used was Whitehaven coal: the quantity 17 pecks, weighing 450 1/2 lbs.

The potatoes being mashed, (without peeling them,) and the greens chopped fine with a sharp shovel, they were mixed together, and 98 lbs of butter, 14 lbs. of onions boiled and chopped fine, 40 lbs. of salt, 1 lb. of black pepper in powder, and 1/2 lb. of ginger, being added, and the whole well mixed together, this food was served out in portions of 1 quart, or about 2 lbs. each, in wooden noggins, holding each 1 quart when full.

Each of these portions of Calecannon (as this food is called in Ireland) served one person for dinner and supper; and each portion cost about 2 1/14 pence, Irish money, or it cost something less than ONE PENNY sterling per pound.

Twelve pence sterling, make thirteen pence Irish.

The expence (reckoned in Irish money) of preparing this food,
was as follows: viz.

L. s. d.

Potatoes, 19 cwt. at 3s. 6d. per cwt. — — 3 6 6
(N.B. They weighed no more than 1615 lbs.
 when picked and washed.)
Greens, 26 flaskets, at 10d. each, — — — 1 1 10
Butter, 98 lbs. at 72s. per cwt. — — — 3 3 0
Onions, 14 lbs. at 2s. per stone, — — — 0 2 0
Ginger, 1/2 lbs. — — — — — — — — 0 1 3
Salt, 40 lbs. — — — — — — — — — 0 1 1
Pepper, 1 lb. — — — — — — — — — 0 1 1

— — — —-

L. 7 16 9

Expence for fuel, 17 pecks of coal,
at 1L. 3s. 3d. per ton, — — — — — 0 3 2 1/2

— — — — — —-

Total L. 7 19 11 1/2

With this kind of food there is no allowance of bread, nor is any necessary.

It would be hardly possible to invent a more nourishing or more palatable kind of food, than Calecannon, as it is made in Ireland; but the expence of it might be considerably diminished, by using less butter in preparing it.

Salted herrings (which do not in general cost much more than a penny the pound) might be used with great advantage to give it a relish, particularly when a small proportion of butter is used.

In this experiment, 273 gallons of water, weighing about 2224 lbs. avoirdupois, and being at the temperature of 55 degrees, was made to boil, (in two hours and 32 minutes,) with the combustion of 346 1/2 lbs. of coal; which gives rather less than 6 1/2 lbs. of water, to each pound of coal consumed; the water being heated 157 degrees, or from 55 to 212 degrees.

According to my experiments, 20 lbs. of water may be heated 180 degrees, (namely from 32 degrees the freezing point, to 212 degrees the temperature of boiling water,) with the heat generated in the combustion of 1 lb. of pine-wood; consequently, the same quantity of wood (1 lb.) would heat 23 lb. of water 157 degrees, or from 55 to 212 degrees.

But M. Lavoisier has shown us by his experiments, that the quantity of heat generated in the combustion of any given weight of coal, is greater than that generated in the combustion of the same weight of dry wood, in the proportion of 1089 to 600; consequently, 1 lb. of coal ought to make 40 3/4 lbs. of water, at the temperature 55 degrees, boil.

But in the foregoing experiment, 1 lb. of coal was consumed in making 6 1/2 lbs. of water boil; consequently, more than 5/6 of the heat generated, or which might with proper management have been generated in the combustion of the coal, was lost, owing to the bad construction of the boiler and of the fire-place.

Had the construction of the boiler and of the fire-place been as perfect as they were in my experiments, a quantity of fuel would have been sufficient, smaller than that actually used, in the proportion of 6 1/2 to 40 3/4, or instead of 450 1/2 lbs. of coal, 71 3/4 lbs. would have done the business; and, instead of costing 3s. 2 1/2d., they would have cost less than 6 1/4 Irish money, or 5 3/4d. sterling, which is only about 1/3 per cent. of the cost of the ingredients used in preparing the food, for the expence of fuel for cooking it.

These computations may serve to show, that I did not exaggerate, when I gave it as my opinion, (in my Essay on Food,) that the expence for the fuel necessary to be employed in cooking ought never to exceed, even in this country, TWO PER CENT. of the value of the ingredients of which the food is composed; that is to say, when kitchen fire-places are well constructed.

Had the ingredients used in this experiment, viz.

> 2234 lbs. of water
> 1615 lbs. of potatoes,
>   98 lbs. of butter,
>   14 lbs. of onions,
>   40 lbs. of salt,
>   1 lb. of pepper, and
>   0 1/2 lb. of ginger,

— — —

making in all 3992 1/2 lbs., been made into a soup, instead of being made into Calecannon, this, at 1 1/4 lb. (equal to one pint and a quarter), the portion would have served to feed 3210 persons.

But if I can show, that in Ireland, where all the coals they burn are imported from England, a good and sufficient meal of victuals for 3210 persons may be provided with the expence of only 5 3/4d. for the fuel necessary to cook it; I trust that the account I ventured to publish in my first Essay, of the expence for fuel in the kitchen of the Military Workhouse at Munich, namely, that it did not amount to so much as 4 1/2d. a day, when 1000 persons were fed, will no longer appear quite so incredible, as it certainly must appear to those who are not aware of the enormous waste which is made of fuel in the various processes in which it is employed.

I shall think myself very fortunate, if what I have done in the prosecution of these my favourite studies, should induce ingenious men to turn their attention to the investigation of a science, hitherto much neglected, and where every new improvement must tend directly and powerfully to increase the comforts and enjoyments of mankind.

END OF THE FIRST VOLUME.

Footnotes for Essay V.

[1] The number of horses in Bavaria alone amount to above 160,000

[2]
A particular account of these military posts is given in the
Second Chapter of the First Essay.

[3] Suffice it to mention one among numberless facts, which might be brought to prove these assertions: The Beggars of our capital carry on an increasing and very lucrative trade, with confessional and communion testimonials, which they sell to people who daringly transgress the holy ecclesiastical laws, by neglecting to confess and receive the holy sacrament of the Lord's Supper at Easter. Some of these impious wretches receive the sacrament, at least twice in a day, in order not to lose their customers; if the demands for communion testimonials are great, or come late. — —Ye priests and preachers of the gospel, can you still forbear raising your voices against Beggars?

[4] To these the President of the Chamber of Finances has since been added.

[5] Since the year 1792 the Elector, to relieve the Institution from that burden, has ordered the police guards to be paid out of the Public Treasury of the Chamber of Finances.

[6] The Bavarian pound which was used in these experiments, and which is divided into 32 loths, is to the pound Avoirdupois as 12,384 is to 10,000, — or nearly as 5 to 4.

This the end of Volume 1 of Count Rumford's Essays

***Etext editors notes follow…
Some modernizations of old spellings have been applied, these are:

show for shew showed for shewed showing for shewing shown for shewn increased for encreased; economical for oeconomical crowded for crouded control for controul

Appendix III contains a table that originally appeared landscape across a number of pages.

I have split this into two, so it will fit comfortable across a normal display screen. I have however added letters to match the two parts together. Also as the concept of pages does not apply, the various 'Carried forward' and 'Brought over totals' have been omitted.